TI-83/84 Plus Graphing Calculators FOR DUMMIES®

PORTABLE EDITION

by C. C. Edwards

WILEY

Wiley Publishing, Inc.

TI-83/84 Plus Graphing Calculators For Dummies®, Portable Edition

Published by
Wiley Publishing, Inc.
111 River Street
Hoboken, NJ 07030-5774

www.wiley.com

Published by Wiley Publishing, Inc., Indianapolis, Indiana

Published simultaneously in Canada

For general information on our other products and services, please contact our Customer Care Department within the U.S. at 800-762-2974, outside the U.S. at 317-572-3993, or fax 317-572-4002.

For technical support, please visit www.wiley.com/techsupport.

Wiley also publishes its books in a variety of electronic formats. Some content that appears in print may not be available in electronic books.

Library of Congress Control Number: 2006925191

ISBN-13: 978-0-470-05610-3

ISBN-10: 0-470-05610-X

Manufactured in the United States of America

10 9 8 7 6 5 4 3 2 1

1B/RW/QV/QW/IN

WILEY

About the Author

C. C. Edwards has a Ph.D. in mathematics from the University of Wisconsin, Milwaukee, and is currently teaching mathematics on the undergraduate and graduate levels. She has been using technology in the classroom since before Texas Instruments came out with their first graphing calculator, and she frequently gives workshops at national and international conferences on using technology in the classroom. She has written forty activities for Texas Instrument's Explorations Web site, and she was an editor of *Eightysomething*, a newsletter formerly published by Texas Instruments. She is also the author of *TI-83 Plus Graphing Calculator For Dummies*.

Just barely five feet tall, CC, as her friends call her, has three goals in life: to be six inches taller, to have naturally curly hair, and to be independently wealthy. As yet, she is nowhere close to meeting any of these goals. When she retires, she plans to become an old lady carpenter.

Publisher's Acknowledgments

We're proud of this book; please send us your comments through our online registration form located at `www.dummies.com/register/`.

Some of the people who helped bring this book to market include the following:

Acquisitions, Editorial, and Media Development

Project Editor: Christopher Morris

Acquisitions Editor: Melody Layne

Copy Editor: Jean Rogers

Technical Editor: Dr. Douglas Shaw, University of Northern Iowa

Editorial Manager: Kevin Kirschner

Media Development Associate Producer: Richard Graves

Editorial Assistant: Amanda Foxworth

Sr. Editorial Assistant: Cherie Case

Cartoons: Rich Tennant (`www.the5thwave.com`)

Composition Services

Project Coordinator: Michael Kruzil

Layout and Graphics: Claudia Bell, Denny Hager, Barbara Moore, Melanee Prendergast, Heather Ryan

Proofreaders: Joe Niesen, Christine Pingleton, Shannon Ramsey

Indexer: Joan Griffitts

Publishing and Editorial for Technology Dummies

Richard Swadley, Vice President and Executive Group Publisher

Andy Cummings, Vice President and Publisher

Mary Bednarek, Executive Acquisitions Director

Mary C. Corder, Editorial Director

Publishing for Consumer Dummies

Diane Graves Steele, Vice President and Publisher

Joyce Pepple, Acquisitions Director

Composition Services

Gerry Fahey, Vice President of Production Services

Debbie Stailey, Director of Composition Services

Contents at a Glance

Table of Contents

Introduction

Do you know how to use the TI-83 Plus or TI-84 Plus family of calculators to do each of the following?

- Create and investigate geometric figures
- Graph functions, inequalities, or transformations of functions
- Create stat plots and analyze statistical data
- Simulate probability experiments like tossing coins, rolling dice, and so on
- Write calculator programs
- Transfer files between two or more calculators
- Save calculator files on your computer
- Add applications to your calculator so that it can do even more than it could when you bought it

If not, then this is the book for you. Contained within these pages are straightforward, easy-to-follow directions that tell how to do everything listed here — and much, much more.

About This Book

Although this book does not tell you how to do *everything* the calculator is capable of doing, it gets pretty close. It covers more than just the basics of using the calculator, paying special attention to warning you of the problems that you could encounter if you knew only the basics of using the calculator.

This is a reference book. It's process-driven, not application-driven. You won't be given a problem to solve and then be told how to use the calculator to solve that particular problem. Instead, you're given the steps needed to get the calculator to perform a particular task, such as constructing a histogram or perpendicular line.

Conventions Used in This Book

When I refer to "the calculator," I am referring to the TI-83 Plus and TI-84 Plus family of calculators because the keystrokes on these calculators are the same. When I want you to press a key on the calculator, I use an icon for that key. For example, if I want you to press the ENTER key, I say press [ENTER]. If I want you to press a series of keys, such as the Stat key and then the right-arrow key, I say (for example) press [STAT][▶]. All keys on the calculator are pressed one at a time. On the calculator, there is no such thing as holding down one key while you press another key.

It's tricky enough to get handy with the location of the keys on the calculator, and even more of a challenge to remember the location of the secondary functions, such as the blue functions on the TI-84 or the yellow functions on the TI-83 that appear above the key. So when I want you to access one of those functions, I give you the actual keystrokes. For example, if I want you to access the Draw menu, I tell you to press [2nd][PRGM]. This is a simpler method than that of the manual that came with your calculator — which would say press [2nd][DRAW] and then make you hunt for the location of the secondary function DRAW. The same principle holds for using key combinations to enter specific characters; for example, I tell you to press [ALPHA][0] to enter a space.

When I want you to use the arrow keys, but not in any specific order, I say press [▶][◀][▲][▼] or use the arrow keys. If I want you to use only the up- and down-arrow keys, I say press [▲][▼].

Foolish Assumptions

My nonfoolish assumption is that you know (in effect) nothing about using the calculator, or you wouldn't be reading this book. My foolish assumptions are as follows:

- You own, or have access to, one of the TI-83 Plus or TI-84 Plus family of calculators.
- If you want to transfer files from your calculator to your computer, I assume that you have a computer and know the basics of how to operate it.

- If you want to write programs on your calculator, I assume that you have some (albeit minor) experience in programming, but not necessarily on the calculator.

Icons Used in This Book

This book uses four icons to help you along the way. Here's what they are and what they mean:

The text following this icon tells you about shortcut and other ways of enhancing your use of the calculator.

The text following this icon tells you something you should remember because if you don't, it may cause you problems later. Usually the Remember icon highlights a reminder to enter the appropriate type of number so that you can avoid an error message.

There is no such thing as crashing the calculator. But this icon warns you of those *few* times when you can do something wrong on the calculator and be totally baffled because the calculator is giving you confusing feedback — either no error message or a cryptic error message that doesn't really tell you the true location of the problem.

This is the stuff you don't need to read unless you're really curious.

Where to Go from Here

This book is designed so that you do not have to read it from cover to cover. You don't even have to start reading at the beginning of a chapter. When you want to know how to get the calculator to do something, just start reading at the beginning of the appropriate section. The Index and Table of Contents should help you find whatever you're looking for. And for your first tip on where to find other information:

See also *TI-83 Plus Graphing Calculator For Dummies* (Wiley) and *TI-84 Plus Graphing Calculator For Dummies* (Wiley), which contain a wealth of useful information.

The 5th Wave
By Rich Tennant
©RICHTENNANT
I'm mathematically dyslexic. But it's not that unusual - 100 out of every 15 people are.

Chapter 1

Coping with the Basics

In This Chapter

- Turning the calculator on and off
- Using the keyboard
- Using the menus
- Setting the mode of the calculator
- Using the Catalog

The TI-83 and TI-84 graphing calculators are loaded with many useful features. With them, you can construct and investigate geometric figures. You can graph and investigate functions, parametric equations, polar equations, and sequences. You can use them to analyze statistical data and to manipulate matrices. You can even use them to calculate mortgage payments.

But if you've never used a graphing calculator before, you may at first find it a bit intimidating. After all, it contains about two dozen menus, many of which contain three or four submenus. But it's really not that hard to get used to using the calculator. After you get familiar with what the calculator is capable of doing, finding the menu that houses the command you need is quite easy. And you have this book to help you along the way.

When to Change the Batteries

The convenience of battery power has a traditional downside: What if the batteries run out of juice at a crucial moment, say during a final exam? Fortunately, the calculator gives you some leeway. When your batteries are low, the calculator displays a

"Your batteries are low" warning message. After you see this message for the first time, the calculator should, according to the manufacturer, continue to function just fine for at least one week. There is one exception: If you attempt to download an application when the batteries are low, the calculator displays a "Batteries are low — Change is required" warning message and refuses to download the application until after you've changed the batteries.

Because you've likely put batteries into countless toys, you should have no trouble opening the cover on the back of the calculator and popping in four AAA batteries. Above the AAA battery chamber is a panel that opens to the compartment containing the backup battery. The type of battery housed in this compartment is indicated on the lid of the panel. The manufacturer recommends that you replace this battery every three or four years. So mark your calendar!

Turning the Calculator On and Off

Press [ON] to turn the calculator on. To turn the calculator off, press [2nd], and then press [ON]. These keys are in the left column of the keyboard. The [ON] key is at the bottom of the column, and the [2nd] key is the second key from the top of this column.

To prolong the life of the batteries, the calculator automatically turns itself off after five minutes of inactivity. But don't worry — when you press [ON], all your work will appear on the calculator just as you left it before the calculator turned itself off.

In some types of light, the screen can be hard to see. To increase the contrast, press and release [2nd] and then hold down [▲] until you have the desired contrast. To decrease the contrast, press [2nd] and hold [▼].

Using the Keyboard

The row of keys under the calculator screen contains the keys you use when graphing. The next three rows, for the most part, contain editing keys, menu keys, and arrow keys. The arrow keys ([▶][◀][▲][▼]) control the movement of the cursor.

The remaining rows contain, among other things, the keys you typically find on a scientific calculator.

Keys on the calculator are *never* pressed simultaneously. In this book, an instruction such as [2nd][ON] indicates that you should first press [2nd] and then press [ON].

Accessing the functions in blue and yellow

Above and to the left of most keys is a secondary key function. These functions are yellow on the TI-83 Plus calculator and blue on the TI-84. To access that function, first press [2nd] and then press the key. For example, π is above the [^] key, so to use π in an expression, press [2nd] and then press [^].

Because hunting for the function can be tedious, in this book I use only the actual keystrokes. For example, I will make statements like, "π is entered into the calculator by pressing [2nd][^]." Most other books would state, "π is entered into the calculator by pressing [2nd] [π]."

When the [2nd] key is active and the calculator is waiting for you to press the next key, the blinking ■ cursor symbol is replaced with the ⇧ symbol.

Using the [ALPHA] key to write words

Above and to the right of most keys is a letter written in green. To access these letters, first press [ALPHA] and then press the key. For example, because the letter O is in green above the [7] key, to enter this letter, press [ALPHA] and then press [7].

Because hunting for letters on the calculator can be tedious, I tell you the exact keystrokes needed to create them. For example, if I want you to enter the letter O, I say, "Press [ALPHA][7] to enter the letter O." Most other books would say "Press [ALPHA] [O]" and leave it up to you to figure out where that letter is on the calculator.

Keys to remember

The following keystrokes are invaluable:

- [2nd][MODE]: This is the equivalent of the Escape key on a computer. It gets you out of whatever you're doing (or have finished doing) and returns you to the Home screen. The Home screen is where the action takes place. This is where you execute commands and evaluate expressions.
- [ENTER]: This key is used to execute commands and to evaluate expressions. When evaluating expressions, it's the equivalent of the equal sign.
- [CLEAR]: This is the "erase" key. If you're entering something into the calculator and change your mind, press this key. If you want to erase the contents of the Home screen, repeatedly press this key until the Home screen is blank.

TIP: You must press [ALPHA] before entering each letter. However, if you want to enter many letters, first press [2nd][ALPHA] to lock the calculator in Alpha mode. Then all you have to do is press the keys for the various letters. When you're finished, press [ALPHA] to take the calculator out of Alpha mode. For example, to enter the word TEST into the calculator, press [2nd][ALPHA][4][SIN][LN][4] and then press [ALPHA] to tell the calculator that you're no longer entering letters.

TIP: When the calculator is in Alpha mode, the blinking ■ cursor symbol is replaced with the Ⓐ symbol. This symbol indicates that the next key you press will insert the green letter above that key. To take the calculator out of Alpha mode, press [ALPHA].

Using the [ENTER] key

The [ENTER] key is used to evaluate expressions and to execute commands. After you have, for example, entered an arithmetic expression (such as 5 + 4), press [ENTER] to evaluate that expression. In this context, the [ENTER] key functions as the equal sign. Entering arithmetic expressions is explained in Chapter 2.

Using the [X,T,Θ,n] key

[X,T,Θ,n] is the key you use to enter the variable in the definition of a function, a parametric equation, a polar equation, or a sequence. In Function mode, this key produces the variable **X**. In Parametric mode it produces the variable **T**; and in Polar and Sequence modes it produces the variables θ and ***n***, respectively. Setting the mode is explained later in this chapter.

Using the arrow keys

The arrow keys ([▶], [◀], [▲], and [▼]) control the movement of the cursor. These keys are in a circular pattern in the upper-right corner of the keyboard. As expected, [▶] moves the cursor to the right, [◀] moves it to the left, and so on. When I want you to use the arrow keys — but not in any specific order — I refer to them all together, as in "Use [▶][◀][▲][▼] to place the cursor on the entry."

What Is the Home Screen?

The Home screen is the screen that appears on the calculator when you first turn it on. This is the screen where most of the action takes place as you use the calculator — it's where you evaluate expressions and execute commands. This is also the screen you usually return to after you've completed a task such as entering a matrix in the Matrix editor or entering data in the Stat List editor.

REMEMBER

Press [2nd][MODE] to return to the Home screen from any other screen. This combination of keystrokes, [2nd][MODE], is the equivalent of the Escape key on a computer. It always takes you back to the Home screen.

REMEMBER

If you want to clear the contents of the Home screen, repeatedly press [CLEAR] until the Home screen is blank.

The Busy Indicator

If you see a moving vertical line in the upper-right corner of the screen, this indicates that the calculator is busy graphing a function, evaluating an expression, or executing a command.

If it's taking too long for the calculator to graph a function, evaluate an expression, or execute a command, and you want to abort the process, press [ON]. If you're then confronted with a menu that asks you to select either **Quit** or **Goto**, select **Quit** to abort the process.

Editing Entries

The calculator offers four ways to edit an entry:

- **Deleting the entire entry:**

  Use [▶][◀][▲][▼] to place the cursor anywhere in the entry and then press [CLEAR] to delete the entry.

- **Erasing part of an entry:**

  To erase a single character, use [▶][◀][▲][▼] to place the cursor on the character you want to delete and then press [DEL] to delete that character.

- **Inserting characters:**

  Because "typing over" is the default mode, to insert characters you must first press [2nd][DEL] to enter Insert mode. When you insert characters, the inserted characters are placed to the left of the cursor. For example, if you want to insert CD between B and E in the word ABEF, you would place the cursor on E to make the insertion.

  To insert characters, use [▶][◀][▲][▼] to place the cursor at the location of the desired insertion, press [2nd][DEL], and then key in the characters you want to insert. When you're finished inserting characters, press one of the arrow keys to take the calculator out of Insert mode.

- **Keying over existing characters:**

  "Type over" is the default mode of the calculator. So if you want to overtype existing characters, just use [▶][◀][▲][▼] to put the cursor where you want to start, and then use the keyboard to enter new characters.

On the Home screen, the calculator doesn't allow you to directly edit an already-evaluated expression or an already-executed command. But you *can* recall that expression or command if it was the last entry you made in the calculator — and when it's recalled, you can edit it. To recall the last expression or command, press [2nd][ENTER]. This makes the calculator paste a copy of the desired expression or command on the Home screen so you can edit it.

Using Menus

Most functions and commands you use are found in the menus housed in the calculator — and just about every chapter in this book refers to them. This section is designed to give you an overview of how to find and select menu items.

Accessing a menu

Each menu has its own key or key combination. For example, to access the Math menu, press [MATH]; to access the Test menu, press [2nd][MATH]. An example of a menu appears in the first picture in Figure 1-1. This is a picture of the Math menu.

Some menus, such as the Math menu, contain submenus. This is also illustrated in the first picture in Figure 1-1. This picture shows that the submenus in the Math menu are MATH, NUM, CPX, and PRB (Math, Number, Complex, and Probability). Repeatedly press [►] to view the items on the other submenus; repeatedly press [◄] to return to the Math MATH submenu. This is illustrated in the second and third pictures in Figure 1-1.

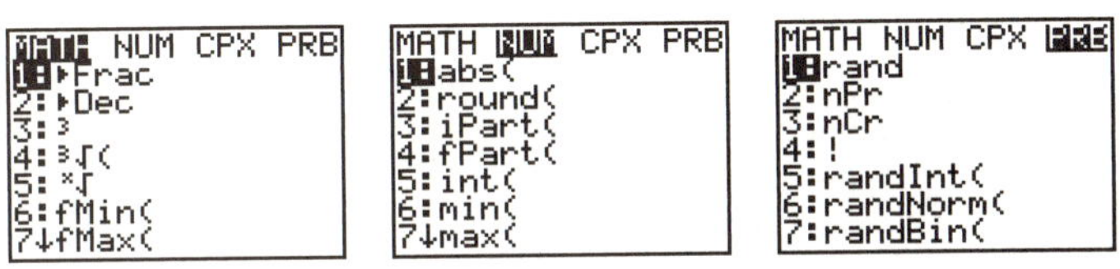

Math MATH menu | Math NUM menu | Math PRB menu

Figure 1-1: Submenus of the Math menu.

Scrolling a menu

After the number 7 in the first two pictures in Figure 1-1, a down arrow indicates that more items are available in the menu than appear on-screen. There's no down arrow after the 7 in the third picture in Figure 1-1 because that menu has exactly seven items.

To see menu items that don't appear on-screen, repeatedly press [▼]. To get quickly to the bottom of a menu from the top of the menu, press [▲]. Similarly, to quickly get from the bottom to the top, press [▼].

Selecting menu items

To select a menu item from a menu, key in the number (or letter) of the item or use [▼] to highlight the number (or letter) of the item and then press [ENTER].

Some menus, such as the Mode menu that is pictured in Figure 1-2, require that you select an item from a list of items by highlighting that item. The list of items usually appears in a single row and the calculator requires that one item in each row be highlighted. To highlight an item, use [►][◄][▲][▼] to place the cursor on the item and then press [ENTER] to highlight the item. The selections on the Mode menu are described in the next section.

Setting the Mode

The Mode menu, which is accessed by pressing [MODE], is the most important menu on the calculator; it tells the calculator how you want numbers and graphs to be displayed. The Mode menu is pictured in Figure 1-2.

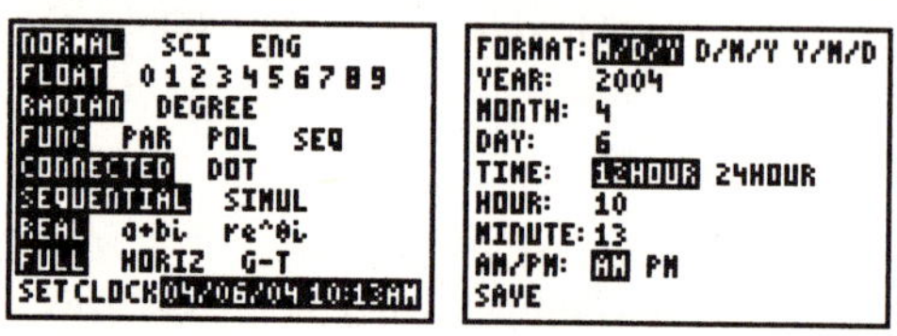

Mode menu Clock menu

Figure 1-2: The Mode and Clock menus.

One item in each row of this menu must be selected. Here are your choices:

- **Normal, Sci, or Eng:**

  This setting controls how numbers are displayed on the calculator. In Normal mode, the calculator displays numbers in the usual numeric fashion that you used in elementary school — provided it can display it using no more than ten digits. If the number requires more than ten digits, the calculator displays it using scientific notation.

In Scientific (**Sci**) mode, numbers are displayed using scientific notation; and in Engineering (**Eng**) mode, numbers are displayed in engineering notation. These three modes are illustrated in Figure 1-3. In this figure, the first answer is displayed in normal notation, the second in scientific notation, and the third in engineering notation.

In scientific and engineering notation, the calculator uses **E*n*** to denote multiplication by 10^n.

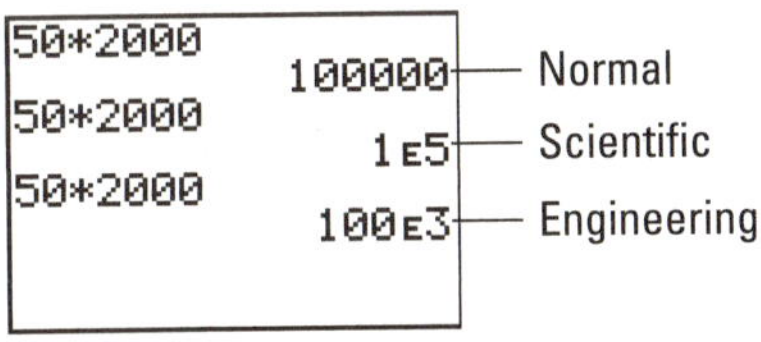

Figure 1-3: Normal, scientific, and engineering notations.

- **Float 0123456789:**

  Select **Float** if you want the calculator to display as many digits as possible. Select **0** if you want all numbers rounded to an integer. If you're dealing with money, select **2** so that all numbers will be rounded to two decimal places. Selecting **5** rounds all numbers to five decimal places, and, well, you get the idea.

- **Radian or Degree:**

  If you select **Radian**, all angles entered in the calculator are interpreted as being in radian measure; all angular answers given by the calculator will also be in radian measure. Similarly, if you select **Degree**, any angle you enter must be in degree measure, and any angular answer given by the calculator is also in degree measure.

- **Func, Par, Pol, or Seq:**

  This setting tells the calculator what type of functions you plan to graph. Select **Func** to graph plain old vanilla functions, $y = f(x)$. Select **Par** to graph parametric equations; **Pol** to graph polar equations; and **Seq** to graph sequences. (Sequences are also called *iterative equations.*)

- **Connected or Dot:**

  In **Dot** mode, the calculator produces a graph by plotting only the points it calculates. In **Connected** mode, the calculator joins consecutively plotted points with a line.

  My recommendation is to select the **Connected** mode because each of the graphing options (**Func**, **Par**, **Pol**, and **Seq**) allows you to select a graphing style, one of which is the dot style.

- **Sequential or Simul:**

  In **Sequential** mode, the calculator completes the graph of one function before it graphs the next function. In Simultaneous (**Simul**) mode, the calculator graphs all functions at the same time. It does so by plotting the values of all functions for one value of the independent variable, and then plotting the values of all functions for the next value of the independent variable.

  Simul mode is useful if you want to see whether two functions intersect at the same value of the independent variable. You have to watch the functions as they are graphed in order to *see* if this happens.

- **Real, a + b*i*, or re^θ*i*:**

  If you're dealing with only real numbers, select the **Real** mode. If you're dealing with complex numbers, select **a + b*i*** if you want the complex numbers displayed in rectangular form. If you want complex numbers displayed in polar form, select the **re^θ*i*** mode.

- **Full, Horiz, or G-T:**

  The **Full** screen mode displays the screen as you see it when you turn the calculator on. The other screen modes are split-screen modes. The **Horiz** mode is for when you want to display a graph and the Y= editor or the Home screen at the same time. Use the **G-T** mode when you want to display a graph and a table at the same time. (The split-screen modes are explained in detail in Chapters 8 and 9.)

- **Set Clock:**

  This is where you set the clock on the calculator. To do this, use the arrow keys to place the cursor on the **SET CLOCK** option and press [ENTER]. You see the second picture in Figure 1-2. You use the arrow keys to move from item to item. To select items in the first, fifth, and eighth rows, place the cursor on the desired item and press

[ENTER] to highlight that item. To enter numbers in the other options, edit the existing number or press [CLEAR] and use the keypad to enter a new number. When you are finished setting the clock, save your settings by placing the cursor on **SAVE** and pressing [ENTER].

REMEMBER

If you're planning on graphing trigonometric functions, put the calculator in Radian mode. Reason: Most trig functions are graphed for $-2\pi \le x \le 2\pi$. That is approximately $-6.28 \le x \le 6.28$. That's not a bad value for the limits on the *x*-axis. But if you graph in Degree mode, you would need $-360 \le x \le 360$ for the limits on the *x*-axis. This is doable . . . but trust me, it's easier to graph in Radian mode.

TIP

If your calculator is in Radian mode and you want to enter an angle in degrees, Chapter 3 tells you how to do so without resetting the mode.

Using the Catalog

The calculator's Catalog houses every command and function used by the calculator. However, it's easier to use the keyboard and the menus to access these commands and functions than it is to use the Catalog. There are several exceptions; for example, the hyperbolic functions are found only in the Catalog. If you have to use the Catalog, here's how to do it:

1. **If necessary, use [▸][◂][▴][▾] to place the cursor at the location where you want to insert a command or function found in the Catalog.**

   The command or function is usually inserted on the Home screen, or in the Y= editor when you're defining a function you plan to graph.

2. **Press [2nd][0] to enter the Catalog.**

   This is illustrated in the first picture in Figure 1-4.

3. **Enter the first letter in the name of the command or function.**

   Notice that the calculator is already in Alpha mode, as is indicated by the ▯ in the upper-right part of the screen. To enter the letter, all you have to do is press the key corresponding to that letter. For example, if you're using the Catalog to access the hyperbolic function tanh, press [4] because the letter **T** is written in green above this key. This is illustrated in the second picture in Figure 1-4.

4. **Repeatedly press [▾] to move the indicator to the desired command or function.**

5. **Press [ENTER] to select the command or function.**

   This is illustrated in the third picture in Figure 1-4. After pressing [ENTER], the command or function is inserted at the cursor location.

Press [2nd] [0] | Enter first letter | Select item

Figure 1-4: Steps for using the Catalog.

Chapter 2

Doing Basic Arithmetic

In This Chapter

- Entering and evaluating arithmetic expressions
- Obeying the order of operations
- Storing and recalling variables
- Combining expressions
- Writing small programs

When you use the calculator to evaluate an arithmetic expression such as $5^{10} + 4^6$, the format in which the calculator displays the answer depends on how you have set the mode of the calculator. Do you want answers displayed in scientific notation? Do you want all numbers rounded to two decimal places?

Setting the *mode* of the calculator affords you the opportunity to tell the calculator how you want these, and other questions, answered. (Setting the mode is explained in Chapter 1.) When you're doing basic, real-number arithmetic, the mode is usually set as it appears in Figure 1-2 of Chapter 1.

Entering and Evaluating Expressions

Arithmetic expressions are evaluated on the Home screen. The Home screen is the screen you see when you turn the calculator on. If the Home screen is not already displayed on the calculator, press [2nd][MODE] to display it. If you want to clear the contents of the Home screen, repeatedly press [CLEAR] until the screen is empty.

Arithmetic expressions are entered in the calculator the same way you would write them on paper if you were restricted to using the division sign (/) for fractional notation. This restriction sometimes requires parentheses around the numerator or the denominator, as illustrated in the first two calculations in Figure 2-1.

There is a major difference between the subtraction [−] key and the [(−)] negation key. They are not the same, nor are they interchangeable. Use the [−] key to indicate subtraction; use the [(−)] key before a number to identify that number as negative. If you improperly use [(−)] to indicate a subtraction problem, or if you improperly use [−] to indicate that a number is negative, you get the ERR: SYNTAX error message. The use of these two symbols is illustrated in the last calculation in Figure 2-1.

When entering numbers, do not use commas. For example, the number 1,000,000 is entered in the calculator as 1000000.

After entering the expression, press [ENTER] to evaluate it. The calculator displays the answer on the right side of the next line, as shown in Figure 2-1.

```
5+3/2
             6.5
(5+3)/2
               4
-3--4
               1
```

Figure 2-1: Evaluating arithmetic expressions.

Important Keys

Starting with the fifth row of the calculator, you find the functions commonly used on a scientific calculator. Here's what they are and how you use them:

- **π and *e*:**

  The transcendental numbers π and *e* are respectively located in the fifth and sixth rows of the last column of the keyboard. To enter π in the calculator, press [2nd][^]; to enter *e*, press [2nd][÷], as shown in the first line of Figure 2-2.

- **The trigonometric and inverse trigonometric functions:**

  The trigonometric and inverse trigonometric functions are located in the fifth row of the keyboard. These functions require that the argument of the function be enclosed in parentheses. To remind you of this, the calculator provides the first parenthesis for you (as in the first line of Figure 2-2).

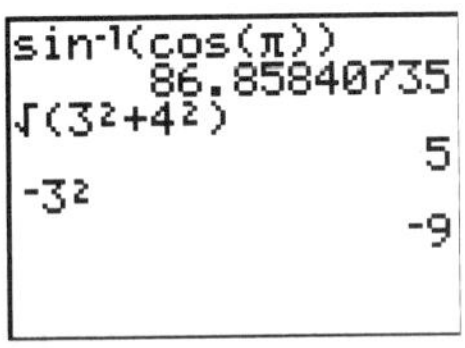

Figure 2-2: Examples of arithmetic expressions.

- **The square-root and exponential functions:**

  The square-root function, 10^x function, and e^x function are respectively located in the sixth, seventh, and eighth rows of the first column on the keyboard. Each of these functions requires that its argument be enclosed in parentheses. To remind you of this, the calculator provides the first parenthesis for you (as in the third line of Figure 2-2).

- **The inverse and square functions:**

  The inverse and square functions are respectively located in the fifth and sixth rows of the left column on the calculator. To enter the multiplicative inverse of a number, enter the number and then press [x^{-1}]. Similarly, to square a number, enter the number and then press [x^2]. The third line of Figure 2-2 shows this operation.

TIP: If you want to evaluate an arithmetic expression and you need a function other than those just listed, you'll most likely find that function in the Math menu (described in detail in the next chapter). The hyperbolic functions are an exception; those you find in the calculator's Catalog. (Chapter 1 discusses the Catalog and how to access the hyperbolic functions.)

Order of Operations

The order in which the calculator performs operations is the standard order that we are all used to. Spelled out in detail, here is the order in which the calculator performs operations:

1. **The calculator simplifies all expressions surrounded by parentheses.**

2. **The calculator evaluates all functions that are followed by the argument.**

   These functions supply the first parenthesis in the pair of parentheses that must surround the argument. An example is sin *x*. When you press [SIN] to access this function, the calculator inserts **sin(** on-screen. You then enter the argument and press [)].

3. **The calculator evaluates all functions entered after the argument.**

   An example of such a function is the square function. You enter the argument and then press [x^2] to square it.

   Evaluating -3^2 may not give you the expected answer. We think of –3 as being a single, negative number. So when we square it, we expect to get +9. But the calculator gets –9 (as indicated in the fifth line of Figure 2-2). This happens because the normal way to enter –3 into the calculator is by pressing [(-)][3] — and pressing the [(-)] key is equivalent to multiplying by –1. Thus, in this context, $-3^2 = -1 * 3^2 = -1 * 9 = -9$. To avoid this potentially hazardous problem, always surround negative numbers with parentheses *before* raising them to a power.

4. **The calculator evaluates powers entered using the [^] key and roots entered using the $\sqrt[x]{\ }$ function.**

   The $\sqrt[x]{\ }$ function is found in the Math menu which is explained in Chapter 3. You can also enter various roots by using fractional exponents — for example, the cube root of 8 can be entered as 8^(1/3).

5. **The calculator evaluates all multiplication and division problems as it encounters them, proceeding from left to right.**

6. **The calculator evaluates all addition and subtraction problems as it encounters them, proceeding from left to right.**

Well, okay, what does the phrase "x plus 1 divided by x minus 2" actually *mean* when you say it aloud? Well, that depends on how you say it. Said without pausing, it means $(x + 1)/(x - 2)$. Said with a subtle pause after the "plus" and another before the "minus," it means $x + (1/x) - 2$. The calculator can't hear speech inflection, so make good use of those parentheses when you're "talking" to the calculator.

Using the Previous Answer

You can use the previous answer in the next arithmetic expression you want to evaluate. If that answer is to appear at the beginning of the arithmetic expression, first key in the operation that is to appear after the answer. The calculator displays **Ans** followed by the operation. Then key in the rest of the arithmetic expression and press [ENTER] to evaluate it, as illustrated in Figure 2-3.

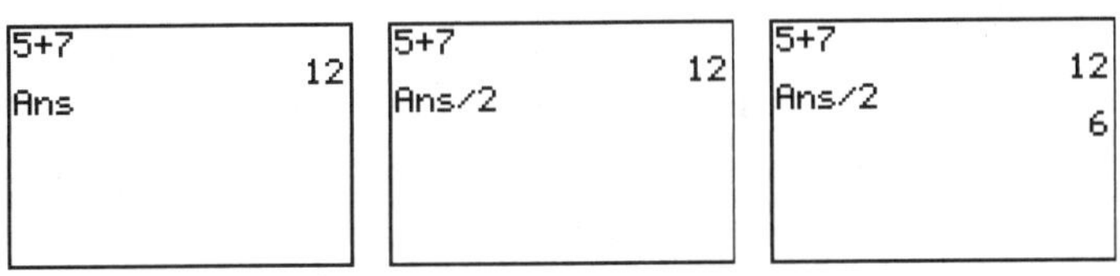

Press [2nd] [(-)] Complete expression Press [ENTER]

Figure 2-3: Steps for starting an expression with the last answer.

If you want to embed the last answer in the next arithmetic expression, key in the beginning of the expression to the point where you want to insert the previous answer. Then press [2nd][(-)] to key in the last answer. Finally, key in the rest of the expression and press [ENTER] to evaluate it, as shown in Figure 2-4.

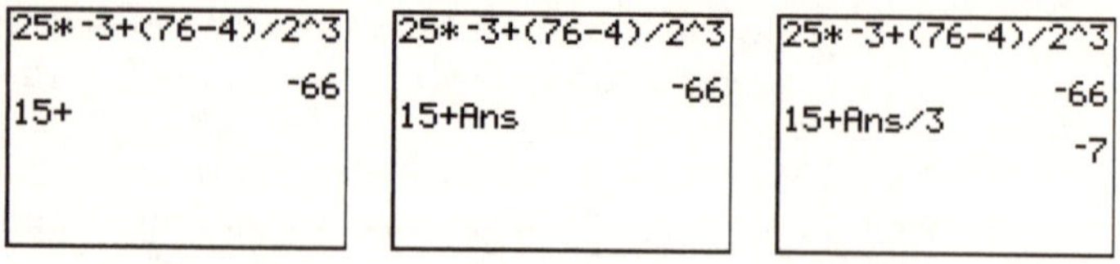

Start expression | Press [2nd] [(-)] | Complete expression

Figure 2-4: Steps for embedding the last answer in an expression.

Recycling the Last Entry

If you want to reuse the last command, function, or expression entered in the calculator — but with different instructions, arguments, or variables — you can simply recall that command, function, or expression and then edit it. To do so, follow these steps:

1. **Enter the command, function, or expression on the Home screen and then press [ENTER] to execute command or evaluate the function.**

   This is illustrated in the first picture in Figure 2-5.

2. **Press [2nd][ENTER] to paste a copy of that command or function on the home screen.**

   This procedure is shown in the second picture in Figure 2-5.

3. **Edit the command, function, or expression and then press [ENTER] to reevaluate it.**

   The third picture in Figure 2-5 shows this procedure. Editing is explained in Chapter 1.

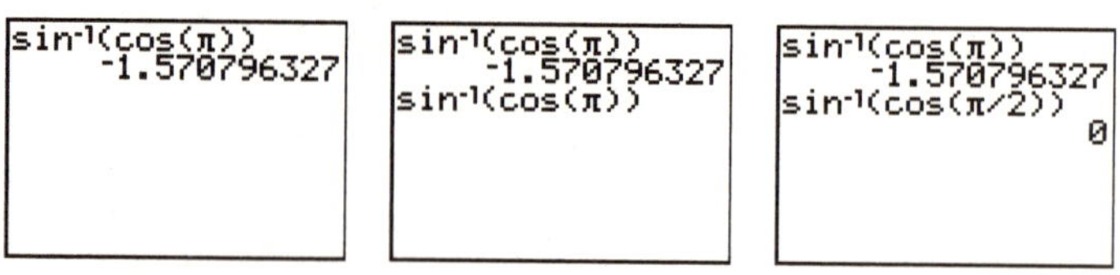

Original expression | Press [2nd] [ENTER] | Edit expression

Figure 2-5: Steps for recycling a command, function, or expression.

Storing Variables

If you plan to use the same number many times when evaluating arithmetic expressions, consider storing that number in a variable. To do so, follow these steps:

1. **If necessary, press [2nd][MODE] to enter the Home screen.**
2. **Enter the number you want to store in a variable.**

   You can store the number as an arithmetic expression. This is illustrated in the first picture in Figure 2-6. After you complete the steps for storing the number, the calculator evaluates that expression.
3. **Press [STO▸].**

   The result of this action is shown in the first picture in Figure 2-6.
4. **Press [ALPHA] and then press the key corresponding to the letter of the variable in which you want to store the number.**

   The second picture in Figure 2-6 shows this process. The letters used for storing variables are the letters of the alphabet and the Greek letter θ.
5. **Press [ENTER] to store the value.**

   This is illustrated in the third picture in Figure 2-6.

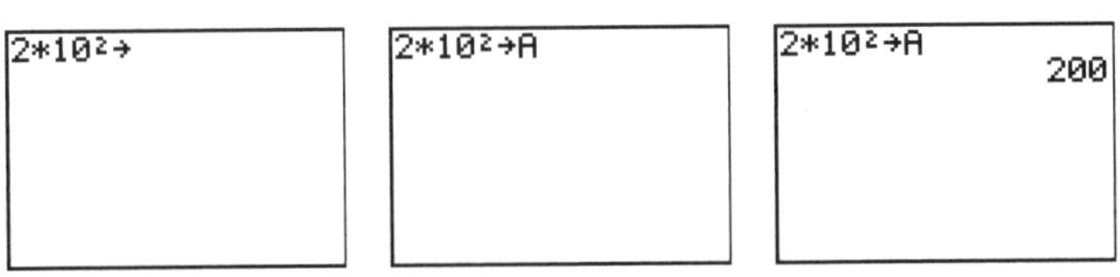

Press [STO▸] Enter variable Press [ENTER]

Figure 2-6: Steps for storing a number in a variable.

TIP: The number you store in a variable remains stored in that variable until you *or the calculator* stores a new number in that variable. Because the calculator uses the letters X, T, and θ when graphing functions, parametric equations, and polar equations, it is possible that the calculator will change the value stored in these variables when the calculator is in graphing mode. For example, if you store a number in the variable X and then ask the calculator to find the zero of the graphed function X^2, the calculator will replace the number stored in X with 0, the zero of X^2. So avoid storing values in these three variables if you want that value to remain stored in that variable after you have graphed functions, parametric equations, or polar equations.

After you have stored a number in a variable, you can insert that number into an arithmetic expression. To do so, place the cursor where you want the number to appear, press [ALPHA], and then press the key corresponding to the letter of the variable in which the number is stored. Figure 2-7 shows how.

Figure 2-7: Inserting a stored variable into an expression.

Combining Expressions

You can *combine* (link) several expressions or commands into one expression by using a colon to separate the expressions or commands. The colon is entered into the calculator by pressing [ALPHA][.].

Combining expressions is a really handy way to write mini-programs, as detailed in the "Writing a mini-program" sidebar.

Writing a mini-program

This figure depicts a program that calculates n^2 when n is a positive integer. The first line of the program initiates the program by storing the value 1 in n and calculating its square. (Storing a number in a variable and combining expressions are explained elsewhere in this chapter.)

The next line on the left shows the real guts of the program. It increments n by 1 and then calculates the square of the new value of n. Of course, the calculator doesn't do these calculations until after you press [ENTER]. But the really neat thing about this is that the calculator will continue to execute this same command each time you press [ENTER]. Because n is incremented by 1 each time you press [ENTER], you get the values of n^2 when n is a positive integer.

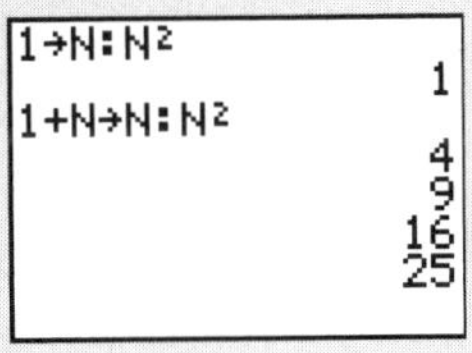

Chapter 3

The Math and Angle Menus

In This Chapter

- Inserting a Math menu function into an expression
- Converting between degrees and radians
- Entering angles in degrees, minutes, and seconds
- Converting between degrees and degrees, minutes, and seconds
- Entering angles in degrees, minutes, and seconds
- Overriding the angle mode of the calculator

Are you hunting for the absolute value function? Look no farther — it's in the Math menu. Do you want to convert a decimal to a fraction? You can find the function that does this is in the Math menu. In general, any math function that cannot be directly accessed using the keyboard is housed in the Math menu. Similarly, any function that deals with angles is housed in the Angle menu. This chapter tells you how to access and use those functions.

The Math Menu and Submenus

Press [MATH] to access the Math menu. This menu contains four submenus: MATH, NUM, CPX, and PRB. Use the [▶][◀] keys to get from one submenu to the next, and back again.

The Math MATH submenu contains the general mathematical functions such as the cubed root function. It also contains the calculator's Equation Solver that, as you would expect, is used to solve equations. The Equation Solver is explained in Chapter 4. The Math NUM submenu contains the functions usually associated with numbers, such as the least common multiple function. A detailed explanation of the functions in these two menus is given later in this chapter.

The Math CPX submenu contains functions normally used with complex numbers. This submenu is explained in detail in the next chapter. The Math PRB submenu contains the probability and random-number functions. (Probability is explained in Chapter 11.)

Using Math menu functions

Most functions housed in the Math menu require that the argument be entered after entering the function. Such functions are easily recognizable by the parenthesis following the name of the function; the cube-root function, $^3\sqrt{}($, is an example.

With one exception, functions in the Math menu that have no parenthesis at the end of their names require that the argument be entered before the function is entered. The cube function is an example of such a function. (The exception is the *x*th root function, $^x\sqrt{}$. This function requires that the root be entered first, then the $^x\sqrt{}$ function, and then the argument.)

Inserting a Math menu function

Math menu functions are usually inserted into arithmetic expressions entered on the Home screen, or into definitions of functions in the Y= editor. Pressing [2nd][MODE] puts you in the Home screen, and pressing [Y=] puts you in the Y= editor. The Y= editor is used to define functions you want to graph. This editor is explained in detail in Chapter 8. To insert a Math menu function into an expression, follow these steps:

1. **If necessary, use [▸][◂][▴][▾] to place the cursor where you want to insert the function.**

   Inserting a Math menu item in an arithmetic expression that is entered on the Home screen is illustrated in Figure 3-1. This figure shows how to insert the cube function in order to evaluate $\sqrt{}(2^3 + 17)$.

2. **Press [MATH].**

3. **Use [▸] to select the appropriate submenu of the Math menu.**

4. **Enter the number of the function or use ▼ to highlight the number of the function and then press ENTER.**

5. **If necessary, complete the expression.**

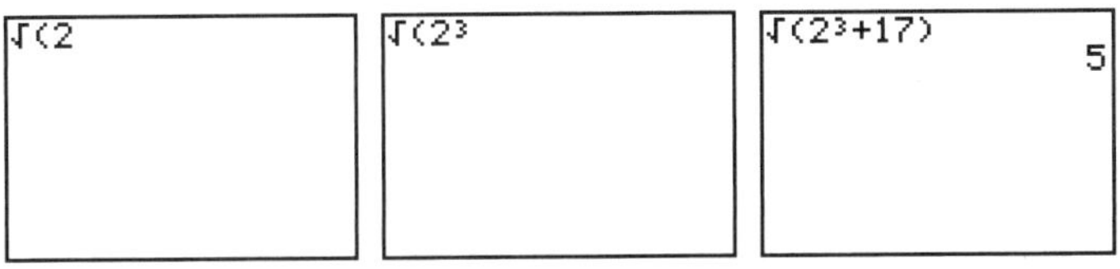

Start expression Insert function Complete expression

Figure 3-1: Steps for inserting a Math menu function in an expression.

The Math MATH submenu

Press MATH to access the Math MATH submenu. This submenu contains general mathematical functions you can insert into an expression. The following sections explain the items housed in this submenu, except for the **Solver** function at the bottom of the Math MATH submenu. This latter function, used to solve equations, is discussed in Chapter 4.

Converting between decimals and fractions

The **Frac** function always converts a finite decimal to a fraction. Sometimes this function can convert an infinite repeating decimal to a fraction. When it can't, it lets you know by redisplaying the decimal. Be sure to enter the decimal before inserting the **Frac** function, as shown in the first two lines of the first picture in Figure 3-2.

The **Dec** function converts a fraction to a decimal. Enter the fraction before you insert the **Dec** function. An example is shown in the first picture in Figure 3-2.

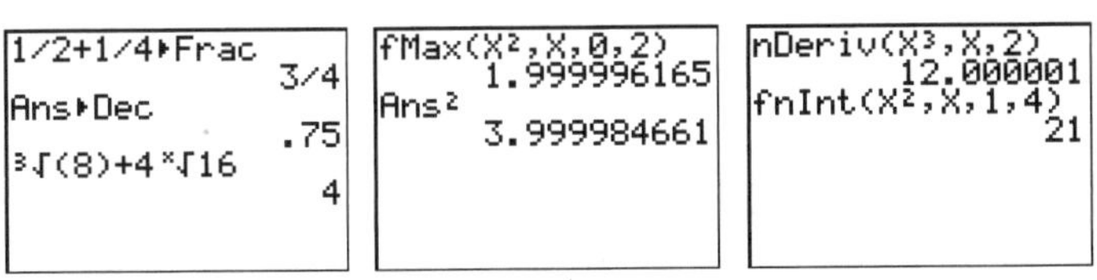

Figure 3-2: Using the functions in the Math MATH submenu.

Cubing and taking cube roots

The cube function, ^{3}, cubes the value that precedes the function (as illustrated in Figure 3-1). The cube-root function, $\sqrt[3]{}$, finds the cube root of a value that follows the function. This value must be enclosed in parentheses; the calculator supplies the opening parenthesis; you must supply the closing parenthesis. The first picture in Figure 3-2 shows an example of using the cube-root function.

Taking the xth root

The xth root function, $\sqrt[x]{}$, finds the xth root of the value that follows the function. To use this function, first enter the root x, then insert the $\sqrt[x]{}$ function, and then enter the argument. This is illustrated in the first picture in Figure 3-2. In this example, the calculator is evaluating the fourth root of 16.

Finding the location of maximum and minimum values

The **fMin** and **fMax** functions approximate *where* the minimum or maximum value of a function occurs in a specified interval. *They do not compute the minimum or maximum value of the function;* they just give you the x-coordinate of the minimum or maximum point. Chapter 10 tells you how to get the calculator to compute minimum and maximum values of a function.

The **fMin** and **fMax** functions are standalone functions in the sense that they cannot be used in an expression. To use these functions, insert the appropriate function, **fMin** or **fMax,** at the beginning of a new line on the Home screen. Then enter the definition of the function whose minimum or maximum you want to locate. Press [,] and enter the variable used in the definition of the function you just entered. Press [,] and enter the lower limit of the specified interval. Then press [,], enter the upper limit, and press [)]. Finally, press [ENTER] to *approximate* the location of the minimum or maximum in the specified interval. This is illustrated in the second picture in Figure 3-2. In this picture the calculator is *approximating* the location of the maximum value of the function x^2 in the interval $0 \leq x \leq 2$.

Doing numerical differentiation and integration

The calculator cannot perform symbolic differentiation and integration. For example, it cannot tell you that the derivative of x^2 is $2x$, nor can it evaluate an indefinite integral. But the **nDeriv** function will approximate the derivative (slope) of a function at a specified value of the variable, and the **fnInt** function will approximate a definite integral.

Insert the **nDeriv** function, enter the function whose derivative you want to find, and then press [,]. Enter the variable used in the definition of the function you just entered and press [,]. Then enter the value at which the derivative is to be taken, and press [)]. Finally, press [ENTER] to *approximate* the derivative. This is illustrated in the third picture in Figure 3-2.

To use the **fnInt** function, insert **fnInt**, and then enter the function you are integrating. Press [,] and enter the variable used in the definition of the function you just entered. Press [,] and enter the lower limit of the definite integral. Press [,], enter the upper limit, and press [)]. Finally, press [ENTER] to *approximate* the definite integral. This is illustrated in the third picture in Figure 3-2.

REMEMBER: The calculator may give you an error message or a false answer if **nDeriv** is used to find the derivative at a nondifferentiable point or if **fnInt** is used to evaluate an improper integral.

The Math NUM submenu

Press [MATH][▶] to access the Math NUM submenu. (Inserting a Math menu function into an expression is explained earlier in this chapter.) The following sections explain the items housed in this submenu.

Finding the absolute value

The **abs** function evaluates the absolute value of the number or arithmetic expression that follows the function. This number or expression must be enclosed in parentheses. The calculator supplies the first parenthesis; you must supply the last parenthesis. An example of using the **abs** function is illustrated in the first picture in Figure 3-3.

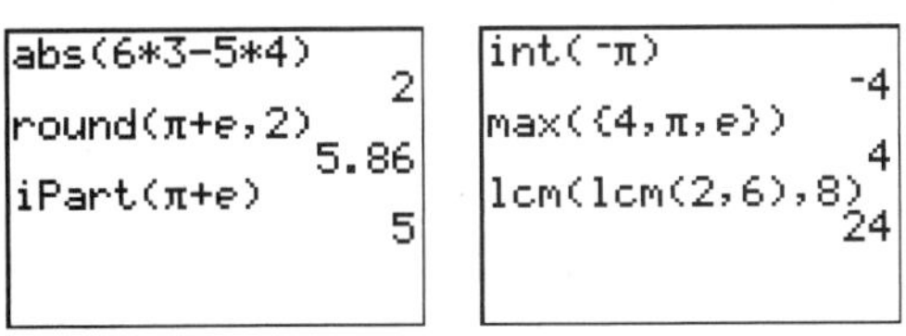

Figure 3-3: Using the functions in the Math NUM submenu.

Rounding numbers

The **round** function rounds a number or arithmetic expression to a specified number of decimal places. The number or expression to be rounded and the specified number of decimal places are placed after the function separated by a comma and surrounded by parentheses. The calculator supplies the opening parenthesis; you must supply the closing parenthesis. An example of using the **round** function is the first picture in Figure 3-3.

Finding the integer and fractional parts of a value

The **iPart** and **fPart** functions (respectively) find the integer and fractional parts of the number, or the arithmetic expression that follows the function. This number or expression must be enclosed in parentheses. The calculator supplies the opening parenthesis; you must supply the closing parenthesis. An example of using the **iPart** function is the first picture in Figure 3-3.

Using the greatest-integer function

The **int** function finds the largest integer that is less than or equal to the number or arithmetic expression that follows the function. This number or expression must be enclosed in parentheses. The calculator supplies the opening parenthesis; you must supply the closing parenthesis. An example of using the **int** function is the second picture in Figure 3-3.

Finding minimum and maximum values in a list of numbers

The **min** and **max** functions find (respectively) the minimum and maximum values in the list of numbers that follows the function. Braces must surround the list, and commas must separate the elements in the list. You can access the braces on the calculator by pressing [2nd][(] and [2nd][)]. The list must be enclosed in parentheses. The calculator supplies the opening parenthesis; you must supply the closing parenthesis. An example of using the **max** function is the second picture in Figure 3-3.

TIP

When using the **min** or **max** function to find the minimum or maximum of a two-element list, you can omit the braces that surround the list. For example, **min**(2, 4) returns the value 2.

Finding least common multiple and greatest common divisor

The **lcm** and **gcd** functions find (respectively) the least common multiple and the greatest common divisor of the two numbers that follow the function. These two numbers must be separated by a comma and surrounded by parentheses. The calculator supplies the opening parenthesis; you must supply the closing parenthesis. An example of using the **lcm** function is the second picture in Figure 3-3.

The Angle Menu

The functions housed in the Angle menu allow you to convert between degrees and radians or convert between rectangular and polar coordinates. They also allow you to convert between degrees and DMS (degrees, minutes, and seconds). You can also override the angle setting in the Mode menu of the calculator when you use these functions. For example, if the calculator is in Radian mode and you want to enter an angle measure in degrees, there is a function in the Angle menu that allows you to do so. (Setting the mode is explained in Chapter 1.)

Converting degrees to radians

To convert degrees to radians, follow these steps:

1. **Put the calculator in Radian mode.**

   Setting the mode is explained in Chapter 1.

2. **If necessary, press [2nd][MODE] to access the Home screen.**

3. **Enter the number of degrees.**

4. **Press [2nd][APPS][1] to paste in the ° function.**

5. **Press [ENTER] to convert the degree measure to radians.**

   This is illustrated at the top of the first picture in Figure 3-4.

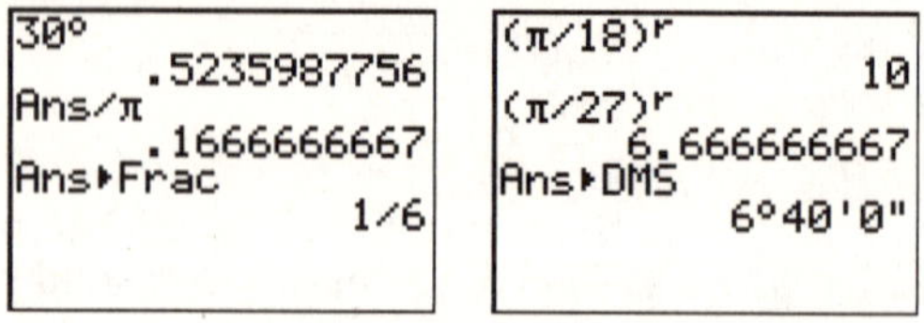

Figure 3-4: Converting between degrees and radians.

TIP

If you're a purist (like me) who likes to see radian measures expressed as a fractional multiple of π whenever possible, continuing with the following steps accomplishes this goal if it's mathematically possible.

6. **Press [÷][2nd][^][ENTER] to divide the radian measure by π.**

   This is illustrated in the third line in the first picture in Figure 3-4.

7. **Press [MATH][1][ENTER] to convert the result to a fraction, if possible.**

   This is illustrated at the bottom of the first picture in Figure 3-4, where 30° is equal to π/6 radians. If the calculator can't convert the decimal obtained in Step 6 to a fraction, it says so by returning the decimal in Step 7.

Converting radians to degrees

To convert radians to degrees:

1. **Put the calculator in Degree mode.**

   Setting the mode is explained in Chapter 1.

2. **If necessary, press [2nd][MODE] to access the Home screen.**

3. **Enter the radian measure.**

   If the radian measure is entered as an arithmetic expression, surround that expression with parentheses. This is illustrated at the top of the second picture in Figure 3-4.

4. **Press [2nd][APPS][3] to paste in the *r* function.**

5. **Press [ENTER] to convert the radian measure to degrees.**

   This is illustrated in the second picture in Figure 3-4.

Converting between degrees and DMS

To convert degrees to DMS (degrees, minutes, and seconds), follow these steps:

1. **Put the calculator in Degree mode.**

   Setting the mode is explained in Chapter 1.

2. **If necessary, press [2nd][MODE] to access the Home screen.**

3. **Enter the degree measure.**

4. **Press [2nd][APPS][4][ENTER] to convert the degrees to DMS.**

   This is illustrated at the bottom of the second picture in Figure 3-4, and also at the top of Figure 3-5.

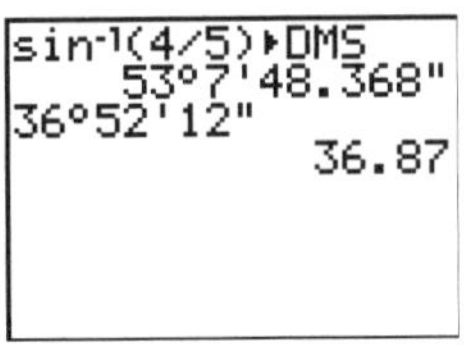

Figure 3-5: Converting between degrees and DMS.

To convert DMS to degrees, follow these steps:

1. **Follow the above Steps 1 and 2.**

2. **Enter the DMS measure.**

   The next section tells you how to insert the symbols for degrees, minutes, and seconds.

3. **Press [ENTER] to convert the DMS entry to degrees.**

   This is illustrated at the bottom of Figure 3-5.

Entering angles in DMS measure

To enter an angle in DMS measure:

1. **Enter the number of degrees and press [2nd][APPS][1] to insert the degree symbol.**

2. **Enter the number of minutes and press [2nd][APPS][2] to insert the symbol for minutes.**
3. **Enter the number of seconds and press [ALPHA][+] to insert the symbol for seconds.**

   This is illustrated at the bottom of Figure 3-5.

Overriding the mode of the angle

If the calculator is in Radian mode but you want to enter an angle in degrees, enter the number of degrees and then press [2nd][APPS][1] to insert the ° degree symbol. If you want to enter an angle in DMS measure while the calculator is in Radian mode, the previous section tells you how to do so. (Setting the mode is explained in Chapter 1.)

If the calculator is in Degree mode and you want to enter an angle in radian measure, enter the number of radians and then press [2nd][APPS][3] to insert the radian-measure symbol. If the radian measure is entered as an arithmetical expression, such as $\pi/4$, be sure to surround it with parentheses before you insert the radian-measure symbol.

Chapter 4

Solving Equations

In This Chapter

- Entering, editing, and solving equations in the Equation Solver
- Assigning values to the variables in the equation
- Defining the interval containing the solution
- Guessing the value of the solution
- Finding multiple solutions

The Equation Solver solves an equation for one variable given the values of the other variables in the equation. The Solver is capable of dealing only with real numbers and is capable of finding only real-number solutions.

Using the Equation Solver

The following lists the basic steps for using the Equation Solver. Each of these steps is explained in detail following this list. If you have never used the Equation Solver before, I suggest that you read the detailed explanations for each step because the Equation Solver is a bit tricky. After you have had experience using the Solver, you can refer back to this list, if necessary, to refresh your memory on its use.

1. **If necessary, set the Mode for real-number arithmetic.**
2. **Enter a new equation (or edit the existing equation) in the Equation Solver.**
3. **Assign values to all variables except the variable you're solving for.**
4. **Enter the bounds for the interval that contains the solution.**

5. Enter a guess for the solution.

6. Press [ALPHA][ENTER] to solve the equation.

Step 1. Set the mode

Because the Equation Solver is equipped to deal only with real numbers, press [MODE] and highlight all entries on the left. (To see what I mean, refer to Figure 1-2 in Chapter 1).

Step 2. Enter or edit the equation to be solved

To enter a new equation in the Equation Solver, follow these steps:

1. **Press [MATH][0] to access the Equation Solver from the Math menu.**

   When the Equation Solver appears, it looks like one of the pictures in Figure 4-1. The first picture shows the Equation Solver when no equation is stored in the Solver; the second picture depicts the Solver and the equation currently stored in it.

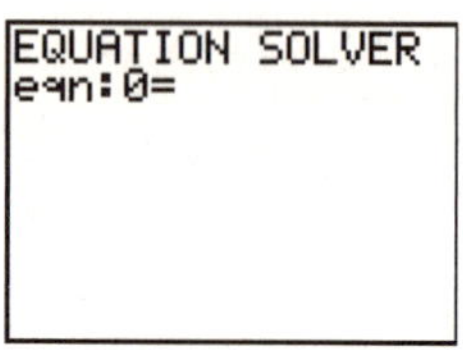

Without an equation

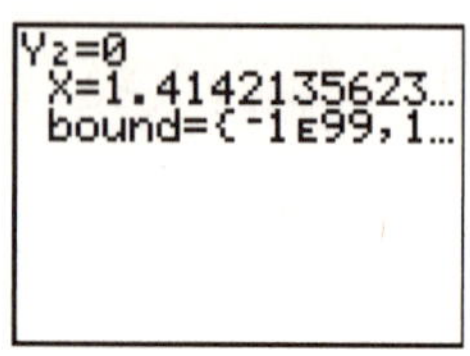

With an equation

Figure 4-1: The Equation Solver.

2. **If your Equation Solver already contains an equation, repeatedly press [▲] until the screen titled EQUATION SOLVER appears.**

   This is illustrated in the first picture in Figure 4-2.

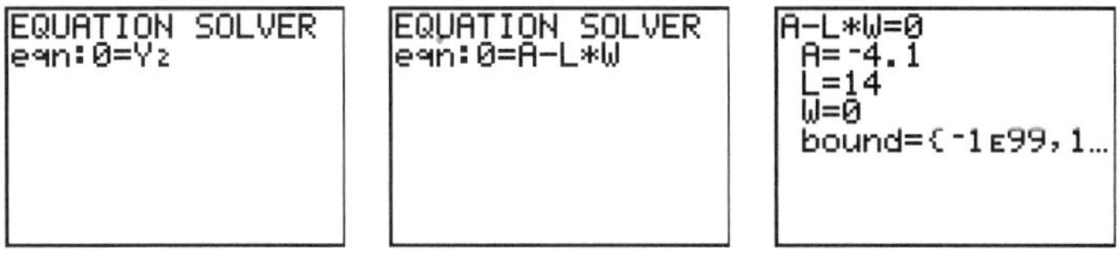

Press [▲] Enter equation Press [ENTER]

Figure 4-2: Steps for entering a new equation in the Equation Solver.

3. **If you need to, press [CLEAR] to erase any equation in the Solver.**
4. **Enter the equation you want to solve.**

   The equation you enter must be set equal to zero and can contain only real numbers. For example, if you want to solve the area equation A = L*W, enter it as 0 = A – L*W or as 0 = L*W – A (as illustrated in the second picture in Figure 4-2).

   You can also use a function that you've entered in the Y= editor in the definition of your equation. For example, if Y_1 = X – 1 in the Y= editor, you can enter $1/Y_1 - 1$ in the Equation Solver to solve the equation $1/(x - 1) - 1 = 0$. To insert such a function into the equation, press [VARS][▶][1] to access the Function menu, and then press the number of the function you want to enter (as in the first picture in Figure 4-2). The Y= editor is explained in Chapter 8.
5. **Press [ENTER] to enter the equation in the solver.**

To edit an equation that is already entered in the Equation Solver, follow these steps:

1. **Follow the above Steps 1 and 2.**
2. **Edit the equation and press [ENTER] when you're finished.**

Step 3. Assign values to variables

After you have entered an equation in the Equation Solver, the values assigned to the variables in your equation are the values that are currently stored in those variables in your calculator. This is illustrated in the last picture in Figure 4-2. You must assign an accurate value to all variables except the variable you're solving for. These values must be real numbers or arithmetic expressions that simplify to real numbers.

To assign a value to a variable, use [▶][◀][▲][▼] to place the cursor on the number currently assigned to that variable and then key in the new value. As you start to key in the new value, the old value is erased. Press [ENTER] when you're finished entering the new value (as illustrated in the first picture of Figure 4-3, where values are assigned to variables **A** and **W**).

Step 4. Define the solution bounds

The **bound** variable at the bottom of the screen, as illustrated in the first picture in Figure 4-3, is where you enter the bounds of the interval containing the solution you're seeking. The default setting for this interval is $[-10^{99}, 10^{99}]$, as is indicated by **bound** = {-1E99, 1E99}. 1E99 is 10^{99} in scientific notation. The ellipsis at the end of the line containing this variable indicates that you have to repeatedly press [▶] to see the rest of the line.

This default setting is more than sufficient for equations that have a unique solution. So if your equation has a unique solution, you don't have to do anything with the value in the **bound** variable.

When the equation you're solving has multiple solutions, it's sometimes necessary to redefine the **bound** variable. Finding multiple solutions is discussed in the last section of this chapter.

To redefine the **bound** variable:

1. **Use [▶][◀][▲][▼] to place the cursor anywhere in the line containing the bound variable.**
2. **Press [CLEAR] to erase the current entry.**
3. **Press [2nd][(] to insert the left brace.**
4. **Enter the lower bound, press [,], enter the upper bound, and then press [2nd][)] to insert the right brace.**
5. **Press [ENTER] to store the new setting in the bound variable.**

Step 5. Guess a solution

Guess at a solution by assigning a value to the variable you're solving for. Any value in the interval defined by the **bound** variable will do. If your guess is close to the solution, the calculator quickly solves the equation; if it's not, it may take the calculator a while to solve the equation. (Assigning a value to a variable in the Equation Solver is explained earlier in this chapter.)

If your equation has more than one solution, the calculator will find the one closest to your guess. The section at the end of this chapter tells you how to find the other solutions.

TIP: If the variable you're solving for is assigned a value (guess) that isn't in the interval defined by the **bound** variable, then you get the ERR: BAD GUESS error message.

Step 6. Solve the equation

To solve an equation, follow these steps:

1. **Use [▶][◀][▲][▼] to place the cursor anywhere in the line that contains the variable you're solving for.**

   This procedure is shown in the second picture in Figure 4-3.

2. **Press [ALPHA][ENTER] to solve the equation.**

   The third picture of Figure 4-3 shows this procedure; the square indicator shown next to the **L** indicates that **L** is the variable just solved for. The **left – rt** value that appears at the bottom of this picture evaluates the two sides of the equation (using the values assigned to the variables) and displays the difference — that is, the accuracy of this solution. A **left – rt** value of zero indicates an exact solution. Figure 4-4 shows a solution that is off by the very small number $-1 * 10^{-11}$.

TIP: If you get the ERR: NO SIGN CHNG error message when you attempt to solve an equation using the Equation Solver, then the equation has no real solutions in the interval defined by the **bound** variable.

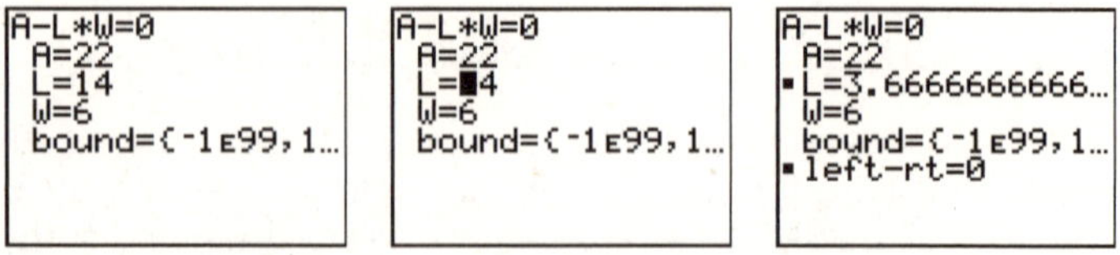

Define variables　　Guess solution　　Press [ALPHA] [ENTER]

Figure 4-3: Steps for solving an equation in the Equation Solver.

Finding Multiple Solutions

To find other solutions to an equation, first find one solution to the equation by following Steps 1 through 6 in the first section of this chapter. This is illustrated in the first picture in Figure 4-4.

Then enter a new guess for the solution you're seeking, or, in the **bound** variable, enter the bounds of an interval that possibly contains a different solution. In the second picture in Figure 4-4 a new guess for the solution is entered.

After making a new guess or after redefining the **bound** variable, follow the steps in the previous section to find another solution to the equation. The third picture in Figure 4-4 shows this procedure.

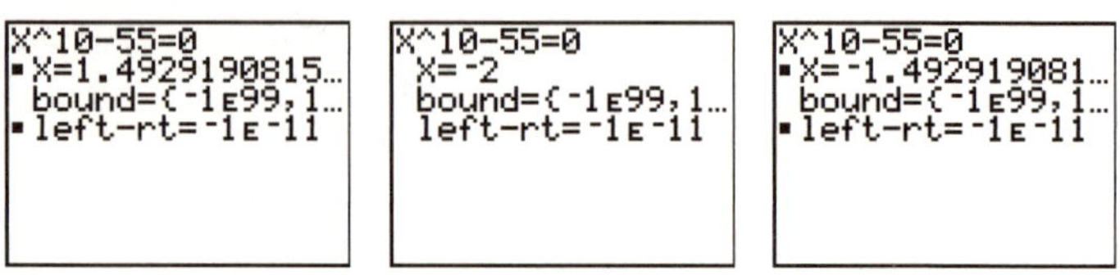

Find 1st solution　　Enter new guess　　Press [ALPHA] [ENTER]

Figure 4-4: Steps for finding multiple solutions to an equation.

Chapter 5

Finding the Best Deal

In This Chapter

- Finding the best interest rate
- Converting between nominal and effective rates
- Deciding whether to lease or to take out a loan
- Finding the internal rate of return

Finding the Best Interest Rate

Which of the following is the best interest rate for a savings account?

- 5.120% annual rate, compounded monthly
- 5.116% annual rate, compounded daily
- 5.115% annual rate, compounded continuously

When *nominal rates* (also called *annual percentage rates*) are compounded at different frequencies (as are those just given), you can compare them to each other only by converting them to *effective rates* (the simple-interest equivalent of nominal rates).

Finding the effective rate

To find the effective rate given the nominal rate:

1. **Set the second line of the Mode menu to Float.**

   When dealing with money, you usually set the second line of the Mode menu to **2** so all numbers are rounded to two decimal places. When you're comparing interest rates, however, you want to see as many decimal places as possible. You can do so by setting the second line to **Float**. (For more about setting the Mode menu, see Chapter 1.)

2. **Press [APPS][1] to start the Finance application.**
3. **Repeatedly press [▼] to move the indicator to the Eff command and press [ENTER].**
4. **Enter the nominal rate, press [,], enter the number of compounding periods per year, and press [)][ENTER].**

   When interest is compounded continuously, it is compounded an infinite number of times a year. But there is no way of entering infinity into the calculator. You can get around this problem by entering a very large number, such as 10^{12}, for the number of compounding periods. The fastest way to enter this number is to press [2nd][,] and then enter the number **12**.

   Figure 5-1 illustrates this procedure; it also shows that the answer to the question posed at the beginning of this section is 5.116%, compounded daily (the choice that gives you the largest effective rate).

You can easily use the same command over and over (as in Figure 5-1) if (after using the command the first time) you press [2nd][ENTER] to recopy the command to the next line on the screen. Then edit the entries in the command and press [ENTER] to execute the command. (Editing is explained in Chapter 1.)

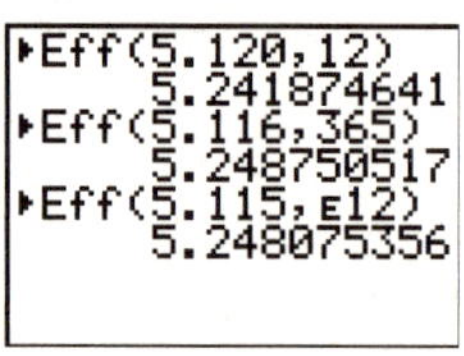

Figure 5-1: Finding the effective rate.

Finding the nominal rate

The steps that convert an effective rate to a nominal rate are similar to those listed in the previous section. To begin, you follow the first two of those steps — and then (in Step 3), select the **Nom** command instead of the **Eff** command. In Step 4, you enter the effective rate (after the comma, enter the number of compounding periods for the nominal rate).

Leasing versus Borrowing

Suppose you're planning to purchase a $2,000 laptop. Which of the following approaches can give you the better deal?

- Lease the laptop for $600 a year for four years with the option to buy the laptop after four years for an additional $300.
- Take out a four-year loan at 12% simple interest.

The *internal rate of return* is the yearly simple interest rate that you earn on an investment plan. In the context of a lease, the internal rate of return is the yearly simple interest rate you would pay if the lease were converted to a loan. So to find the better deal, you must compare the lease's internal rate of return to 12% (the loan's internal rate of return).

To find the lease's internal rate of return:

1. **Select the** irr **command from the Finance application menu.**

   To do so, follow the first two steps in the previous section ("Finding the Effective Rate") — but in Step 3, select the **irr** command instead of the **Eff** command.

2. **Enter the initial cash flow and press [.].**

   For the leasing program described at the beginning of this section (for example), the initial cash flow is the $2,000 price of the laptop. The sidebar at the end of this chapter explains why this value is positive even though the cash is flowing away from you.

3. **Enter the cash-flow list and press [,].**

   In the leasing-program scenario, the cash flow indicates what you paid per year to lease the laptop: $600 a year for the first three years, and $900 in the fourth year (that is, $600 to lease it and then $300 to purchase it). So the cash-flow list is {–600, –900}. (Don't worry: The sidebar at the end of this chapter explains why these values must be negative.) You enter the frequencies for this list in the next step.

   Enter the cash-flow list as a list contained within braces, using commas to separate the elements in the list (as in Figure 5-2). You enter the braces into the calculator by pressing [2nd][(] and [2nd][)]. (Remember to use the [(-)] key to indicate that a number is negative.)

4. **Enter the cash-flow frequency list.**

   The *cash-flow frequency list* indicates how frequently each element (in this case, each specific amount) occurs in the cash-flow list. In the leasing-program scenario, the cash-flow frequency list is {3, 1} because $600 was paid for the first three years and $900 was paid in the fourth year. Figure 5-2 shows this case.

5. **Press [)][ENTER] to calculate the internal rate of return.**

   This procedure is illustrated in Figure 5-2. This figure also shows the answer to the question posed at the beginning of this section: You're better off taking out the loan at 12% simple interest because leasing the laptop is equivalent to a 12.29% loan.

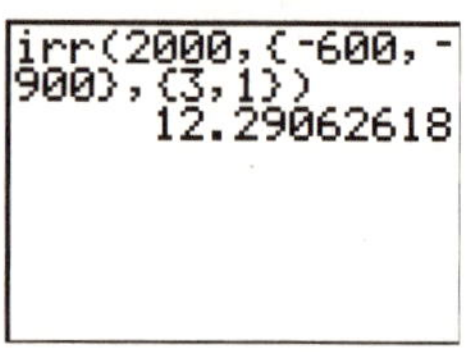

Figure 5-2: Finding the internal rate of return.

When is cash flow negative?

Cash flow is the money that changes hands. The calculator does not know who is involved in this exchange, so it requires that you indicate which way the money is going by using a positive or negative value for cash flow. Figuring out which sign you should use isn't always easy.

The best way to figure out which sign to use is to ask yourself whether the money is going into your pocket or coming out of your pocket. If it is going *into* your pocket, then you have more, so the cash flow is positive. If it is coming *out of* your pocket, then you have less, so it's negative.

For example, if you put $2,000 in a savings account, that's a negative cash flow because it came out of your pocket and went to the bank. It's still your money, but it's no longer in your pocket.

On the other hand, if you take out a $2,000 loan, then that's a positive cash flow because the bank has given you the money. In your mind, it's negative because you have to pay it off. But to the calculator, it's positive because now you have the money.

Chapter 6

Loans and Mortgages

In This Chapter

- Using the Time-Value-of-Money (TVM) Solver
- Using a TVM value in a calculation
- Finding the principal paid on a loan during a specified time period
- Finding the interest paid on a loan during a specified time period
- Finding the balance of a loan after a specified time period

You have a 30-year, $200,000 mortgage on your house. The mortgage rate is 7%, compounded monthly.

- What are your monthly mortgage payments?
- What is the total cost of the loan?
- How much of your first payment is devoted to paying off the balance of the loan?
- How much of the loan was paid off during the second year?
- How much interest do you pay during the life of the loan?
- How much do you still owe on the house after 20 years?

This chapter shows you how to get the calculator to answer these and other, similar questions.

Using the TVM Solver

The **TVM** (time-value-of-money) **Solver** can be used to answer questions like those posed at the beginning of this chapter. In fact, if you tell the **TVM Solver** any four of the following five variables, it will figure out the fifth variable for you:

- **N:** Total number of payments
- **%:** Annual interest rate
- **PV:** Present value
- **PMT:** Amount of each payment
- **FV:** Future value

To get the calculator to find these answers, follow these steps:

1. **Set the second line in the Mode menu to 2.**

   This setting makes the calculator round all numbers to two decimal places, the standard format for money. (Setting the Mode menu is described in Chapter 1.)

2. **Press [APPS][1][1] to select the TVM Solver from the Finance application menu.**

3. **Enter values for four of the first five variables listed in the TVM Solver. Press [ENTER] after making each entry.**

   Some values that are entered in the **TVM Solver** must be entered as negative numbers. For an explanation of when you have to do this, see Chapter 5.

   This step is illustrated in the first picture in Figure 6-1. In this figure, the **TVM Solver** is set up to solve the first question asked at the beginning of this chapter.

   Don't worry about any value currently assigned by the calculator to the variable that the **TVM Solver** is going to find for you. In this example, that variable is **PMT**, the monthly payment.

   For a loan, the present value is always the amount of the loan; the future value, after the loan is paid off, is (naturally) 0.

   You can enter arithmetic problems as values in the **TVM Solver**. The calculator will do the arithmetic after you press [ENTER]. For example, in the first picture in Figure 6-1, **N** was entered as 30*12.

4. **Enter values for P/Y and C/Y. Press [ENTER] after making each entry.**

   - **P/Y** is the number of payments made each year.
   - **C/Y** is the number of times interest is compounded each year.

The calculator assumes these two values are the same. So after you enter the value for **P/Y** and press [ENTER], the same value is assigned to the variable **C/Y**. (If the values are *not* the same in the scenario for which you're using the **TVM Solver**, enter the correct value for **C/Y**. As an example, if you're making monthly payments but the interest rate is a simple interest rate, then **P/Y** = 12 and **C/Y** = 1.)

If interest is compounded continuously, it is compounded an infinite number of times a year. But there is no way of entering infinity into the calculator. You can get around this problem by setting **C/Y** equal to a large number, such as 10^{12}.

5. **In the last line, indicate whether payments are made at the beginning or the end of the payment period.**

   To do so, use [▶][◀] to place the cursor on the appropriate entry, either **END** or **BEGIN**, and press [ENTER].

6. **Place the cursor on the variable for which you want to solve.**

   This operation is shown in the second picture in Figure 6-1. In this figure the calculator is told to solve for **PMT**, the monthly payment.

7. **Press [ALPHA][ENTER] to solve for the variable.**

   The calculator places a square next to the variable it just found. This is illustrated in the third picture in Figure 6-1.

   See the sidebar in Chapter 5 for an explanation of why **PMT** is negative in the last picture in Figure 6-1.

8. Press yz to exit (quit) the TVM Solver and return to the Home screen.

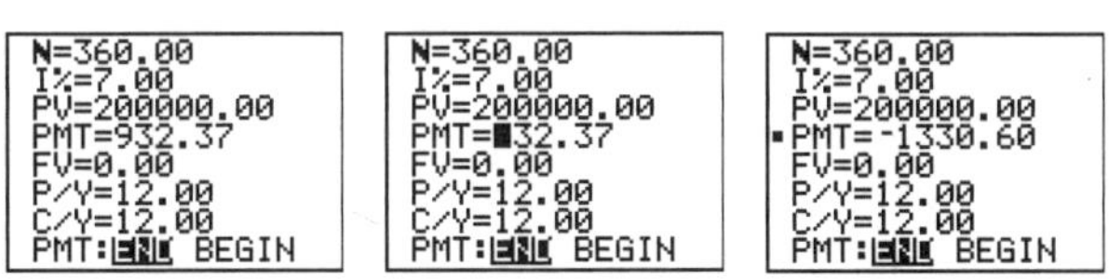

Define variables | Select unknown variable | Press [ALPHA] [ENTER]

Figure 6-1: Steps for using the TVM Solver.

Using a TVM Value

After you've assigned values to all variables in the **TVM Solver** (as explained in the previous section), you can paste these values into other expressions by pressing [APPS][1][▸] and entering the number of the variable you want to use.

An example of this use appears in Figure 6-2. Also, this figure answers the question (*What is the total cost of the loan?*) posed at the beginning of this chapter. Notice that this calculation was done on the Home screen. To get from the **TVM Solver** to the Home screen, press [2nd][MODE].

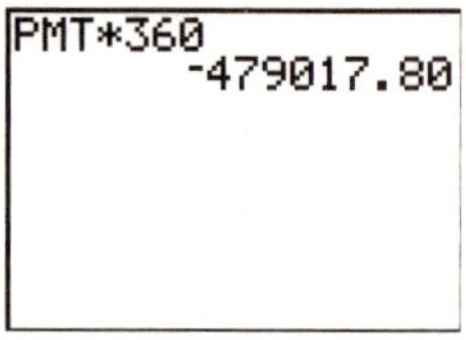

Figure 6-2: Using a TVM variable in a calculation.

Finding Principal and Interest

Use the **ΣPrn** and **ΣInt** commands to compute the principal and interest paid on a loan during a specified time period. To use these commands, follow these steps:

1. **Assign values to all variables in the TVM Solver.**

   This chapter's "Using the TVM Solver" section tells you how.

2. **Press [2nd][MODE] to enter the Home screen.**
3. **Press [APPS][1] to access the Finance application menu.**
4. **Repeatedly press [▾] to select ΣPrn or ΣInt and press [ENTER].**
5. **Enter the number of the first payment in the specified time period, press [,], enter the number of the last payment in the time period, press [)], and then press [ENTER] to display the result.**

This process is illustrated in Figure 6-3 (where you can also see answers to the third, fourth, and fifth questions posed at the beginning of this chapter).

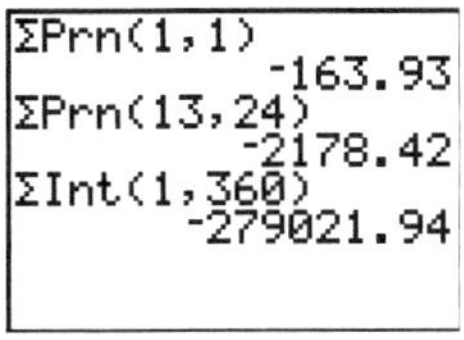

Figure 6-3: Evaluating cumulative principal and interest payments.

Finding the Balance

Use the **bal** command to compute the balance of a loan after a specified time period. To use this command, follow these steps:

1. **Follow Steps 1 through 3 in the previous section.**
2. **Repeatedly press [▼] to select the** bal **command and press [ENTER].**
3. **Enter the number of payments in the specified time period.**

   You may find the "How Many Days Till Christmas?" sidebar in the next chapter useful for calculating the number of payments in a specified time period.

   You can enter the number of payments in the **bal** command as an arithmetic expression, as in Figure 6-4.
4. **Press [)][ENTER] to display the result.**

   This procedure is shown in Figure 6-4. (This figure answers the last question posed at the beginning of this chapter.)

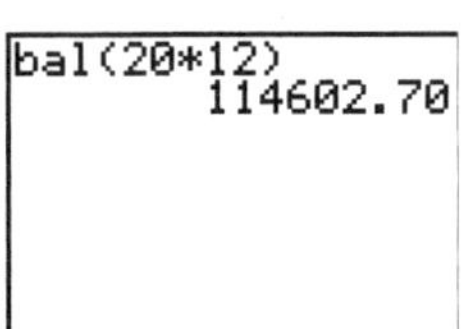

Figure 6-4: Finding the balance of a loan.

Understanding round-off errors

Do you see a problem with the results the calculator gave in Figures 6-2 and 6-3? According to Figure 6-2, the total cost of the house in the scenario posed at the beginning of this chapter is $479,017.80. According to Figure 6-3, however, the total cost is $479,021.94 (total interest plus the $200,000 mortgage). That's a difference of $4.14! The differences in these calculations are caused by 30 years worth of *round-off errors.*

In Figure 6-1, the calculator rounded the monthly payments to $1,330.60. But if you had put the calculator in Float mode, the calculator would have told you that the monthly payments are $1,330.60499. When the calculator uses this value to evaluate other entities, it uses an even better approximation than this one. But it's still using an approximation, so expect round-off errors.

The mortgage lender would avoid this problem by charging $1,330.61 a month for all but the last month.

Chapter 7

Savings and Investments

In This Chapter

- Finding how long it takes to reach a financial goal
- Finding the future value of money
- Finding the present value of money
- Calculating the number of days between two dates

Say you're investing money in an account that earns 6% interest, compounded continuously.

- If you invest $50,000 now and then add $1,000 to the account each month, how many years will it take for the account to reach $1,000,000?
- If you invest $50,000 now and then add $1,000 to the account each month, how much money will you have after 30 years?
- How much should you invest now so that the account will be worth at least $1,000,000 in 30 years?

The **TVM** (time-value-of-money) **Solver** can be used to answer questions like those posed above.

Reaching Financial Goals

You can use the **TVM Solver** to determine how long to leave money in a savings or investment account to reach a specified financial goal. Chapter 6 provides the steps for using the **TVM Solver**; here are some pointers you may want to use as you follow those steps:

- **N** (number of payments):

  This is the variable you're solving for. So in Step 4, assign values to all variables except **N**.

- **PMT** (amount of payment) and **P/Y** (payments per year):

  If you put money into an account and just leave it there without adding to it, then **PMT** is 0. And because you're not making any payments, you can set **P/Y** to 1. This will tell the calculator to evaluate **N** in years.

  On the other hand, if you add to the account on a regular basis, then **PMT** is the value of these regular additions to the account and **P/Y** is the number of times a year these additions to the account are made. This is illustrated in the first picture in Figure 7-1. In this picture, the **TVM Solver** is solving the first question posed at the beginning of this chapter.

- After you have solved for **N** in Step 8:

  If **P/Y** is 1, then the solution for **N** is in years. If **P/Y** isn't 1, then **N** is the total number of payments made. Because **P/Y** payments are made each year, the number of years it takes to reach your financial goal is **N** divided by **P/Y**. Chapter 6 explains how to use a **TVM** value, in particular, how to use these variables in a calculation (as shown in the second picture in Figure 7-1).

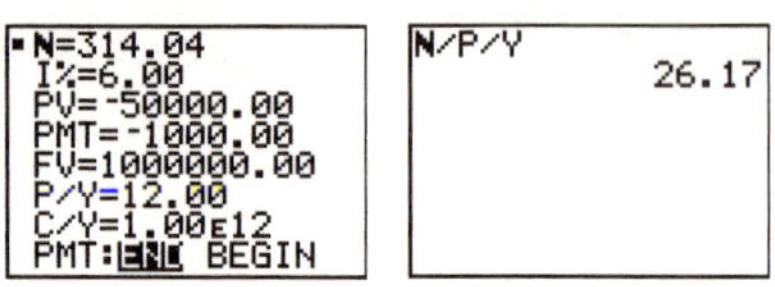

Finding **N** Converting **N** to years

Figure 7-1: Solving for the time needed to reach a financial goal.

Finding Future Value of Money

In the context of a savings or investment account, the future value of your money is the amount of money in the account after a specified time period. The second question posed at the beginning of this chapter is asking for the future value of the account.

To find the future value, follow the steps given in Chapter 6 for using the TVM Solver. Here are some pointers you may want to use:

- **FV** (future value):

  This is the variable you're solving for. So in Step 4, you have to assign values to all variables except **FV**.

- **PMT** (amount of payment) and **P/Y** (payments per year):

  If necessary, read the comments made about **PMT** and **P/Y** in the previous section.

- **N** (number of payments):

  N is the number of years you have the account times **P/Y**.

Figure 7-2 illustrates using the **TVM Solver** to answer the second question posed at the beginning of this chapter.

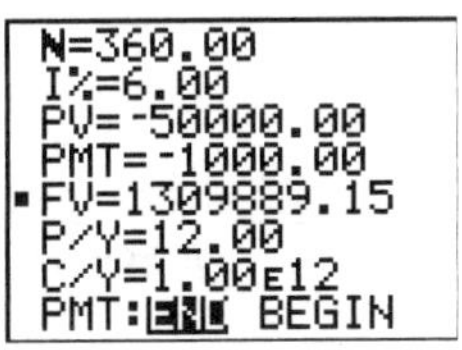

Figure 7-2: Finding the future value of money.

Finding Present Value of Money

In the context of a savings or investment account, the *present value* of money is the amount needed to generate a specific amount of money after a specified time period. The third question posed at the beginning of this chapter asks for the present value of an account.

To find the present value, follow the steps given in Chapter 6 for using the TVM Solver. Here are some pointers you may want to use:

- **PV** (present value):

  This is the variable you're solving for. Accordingly, in Step 4, assign values to all variables except **PV**.

- **FV** (future value):

  This is the amount of money you want to have in the account after the specified time period.

- **PMT** (amount of payment) and **P/Y** (payments per year):

  If necessary, read the comments made about **PMT** and **P/Y** in the section titled "Reaching Financial Goals."

- **N** (number of payments):

  N is the number of years you have the account times **P/Y**.

How many days till Christmas?

To calculate the number of days between two dates:

1. **Press [APPS][1][ALPHA][x^{-1}] to select the dbd command from the Finance application menu.**

2. **Enter the first date using either the *mm.ddyy* or the *ddmm.yy* format.**

   In both formats, ***mm*** is the two-digit number of the month, *dd* is the two-digit number of the day of the month, and *yy* is the last two digits of the year. The date must fall somewhere in the range of years from 1950 through 2049.

   In the U.S.-format *mm.ddyy*, *mm* and *ddyy* are separated by a period; there is no separation between *dd* and *yy*. For example, March 5, 2004, is entered as 03.0504. In the European-format *ddmm.yy*, this date is entered as 0503.04.

3. **Press [,], enter the second date, and then press [)].**

4. **Press [ENTER] to display the number of days between the two dates.**

Using the U.S. date format, the figure given here shows that there are 295 days between March 5, 2004, and Christmas of the same year.

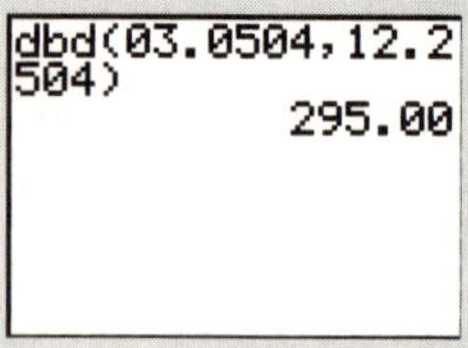

Figure 7-3 illustrates using the **TVM Solver** to answer the third question posed at the beginning of this chapter.

Figure 7-3: Finding the present value of money.

Chapter 8

Graphing Functions

In This Chapter

- Entering functions into the calculator
- Graphing functions
- Recognizing whether the graph is accurate
- Graphing piecewise-defined and trigonometric functions
- Viewing the graph and the function on the same screen
- Saving and recalling a graph and its settings in a Graph Database

The calculator has a variety of features that help you painlessly graph a function. The first step is to enter the function into the calculator. Then to graph the function, you set the viewing window and press [GRAPH]. Or (better yet) you can use one of several Zoom commands to get the calculator to set the viewing window for you. Finally, after you have graphed the function, you can use Zoom commands to change the look of the graph. For example, you can zoom in or zoom out on a graph the same way that a zoom lens on a camera lets you zoom in or out on the subject of the picture you're about to take.

Entering Functions

Before you can graph a function, you must enter it into the calculator. The calculator can handle up to ten functions at once, $\mathbf{Y_1}$ through $\mathbf{Y_9}$ and $\mathbf{Y_0}$. To enter functions in the calculator, perform the following steps:

1. **Press [MODE] and put the calculator in Function mode, as shown in Figure 8-1.**

To highlight an item in the Mode menu, use the [▶][◀][▲][▼] keys to place the cursor on the item, and then press [ENTER]. Highlight **Func** in the fourth line to put the calculator in Function mode. (For more about the other items on the Mode menu, refer to Chapter 1.)

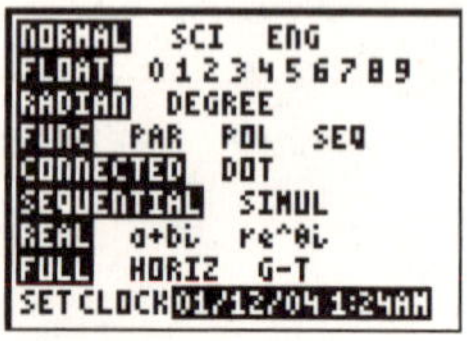

Figure 8-1: Setting Function mode.

2. **Press [Y=] to access the Y= editor.**
3. **Enter the definitions of your functions.**

   To erase an entry that appears after $\mathbf{Y_n}$, use the [▶][◀][▲][▼] keys to place the cursor to the right of the equal sign and press [CLEAR]. Then enter your definition for the new function and press [ENTER].

When you're defining functions, the only symbol the calculator allows for the independent variable is the letter X. Press [X,T,Θ,*n*] to enter this letter in the definition of your function. In Figure 8-2, this key was used to enter the functions $\mathbf{Y_1}$, $\mathbf{Y_2}$, $\mathbf{Y_4}$, $\mathbf{Y_5}$, and $\mathbf{Y_6}$.

As a timesaver, when entering functions in the Y= editor, you can reference another function in its definition. (Figure 8-2, for example, shows function $\mathbf{Y_3}$ defined as $-Y_2$.) To paste a function name in the function you're entering in the Y= editor, follow these steps:

1. **Press [VARS][▶] to access the Y-Variables menu.**
2. **Press [1] to access the Function menu.**
3. **Press the number key for the name of the function you want to paste in the definition.**

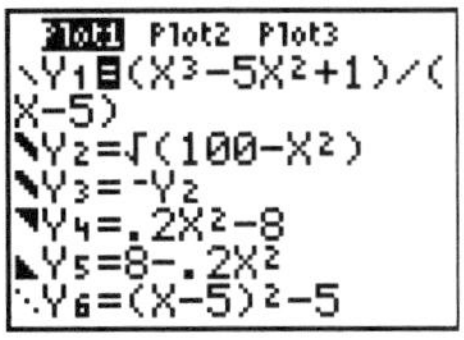

Figure 8-2: Examples of entering functions.

Graphing Functions

Here's where your calculator draws pretty pictures. After you have entered the functions into the calculator, as described in the previous section, you can use the following steps to graph the functions:

1. **Turn off any Stat Plots that you don't want to appear in the graph of your functions.**

   The first line in the Y= editor tells you the graphing status of the Stat Plots. If **Plot1**, **Plot2**, or **Plot3** is highlighted, then that Stat Plot will be graphed along with the graph of your functions. If it's not highlighted, it won't be graphed. In Figure 8-2, **Plot1** will be graphed along with the functions.

   To turn off a highlighted Stat Plot in the Y= editor, use the [▶][◀][▲][▼] keys to place the cursor on the highlighted Stat Plot and then press [ENTER]. The same process is used to highlight the Stat Plot again in order to graph it at a later time.

   When you're graphing functions, Stat Plots can cause problems if they're turned on when you don't really want them to be graphed. The most common symptom of this problem is the ERR: INVALID DIM error message — which by itself gives you almost no insight into what's causing the problem. So if you aren't planning to graph a Stat Plot along with your functions, make sure all Stat Plots are turned off.

2. **Press [2nd][ZOOM] to access the Format menu.**

3. **Set the format for the graph by using the [▶][◀][▲][▼] keys to place the cursor on the desired format, and then press [ENTER] to highlight it.**

In the Format menu, each line of the menu will have one item highlighted. An explanation of each menu selection follows:

- **RectGC and PolarGC:** This gives you a choice between having the coordinates of the location of the cursor displayed in (*x*, *y*) rectangular form or in (*r*, θ) polar form. Select **RectGC** for rectangular form or **PolarGC** for polar form.
- **CoordOn and CoordOff:** This tells the calculator whether you want to see the coordinates of the cursor location displayed at the bottom of the screen as you move the cursor. Select **CoordOn** if you want to see these coordinates; select **CoordOff** if you don't.
- **GridOff and GridOn:** If you select **GridOn**, grid points appear in the graph at the intersections of the tick marks on the *x*- and *y*-axes (as illustrated in Figure 8-3). If you select **GridOff**, no grid points appear in the graph.
- **AxesOn and AxesOff:** If you want to see the *x*- and *y*-axes on your graph, select **AxesOn**. If you don't want to see them, select **AxesOff**.
- **LabelOff and LabelOn:** If you want the *x*- and *y*-axes to be labeled, select **LabelOn** (as in Figure 8-3). Because the location of the labels isn't ideal, selecting **LabelOff** is usually a wise choice.

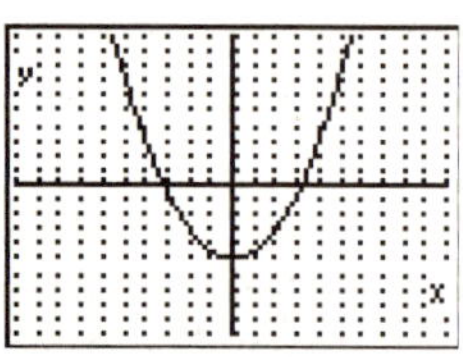

Figure 8-3: A graph with grid points and labeled axes.

- **ExprOn and ExprOff:** If you select **ExprOn**, when you're tracing the graph of a function, the definition of that function appears in the upper left of the screen. If you select **ExprOff** and **CoordOn**, then only the number of the function appears when you trace the function. If you select **ExprOff** and **CoordOff**, then nothing at all appears on the screen to indicate which function you're tracing.

4. **Press [WINDOW] to access the Window editor.**

5. **After each of the window variables, enter a numerical value that is appropriate for the functions you're graphing. Press [ENTER] after entering each number.**

   Figure 8-4 shows the Window editor when the calculator is in Function mode. The items in this menu determine the viewing window for your graph — in particular, how the *x*- and *y*-axes look on the screen. The following gives an explanation of the variables you must set in this editor:

   - **Xmin and Xmax:** These are, respectively, the smallest and largest values of *x* in view on the *x*-axis.

     If you don't know what values your graph will need for **Xmin** and **Xmax**, press [ZOOM][6] to invoke the **ZStandard** command. This command automatically graphs your functions in the Standard viewing window; the settings for this window appear in Figure 8-4. You can then, if necessary, use the other Zoom commands (described in Chapter 9) to get a better picture of your graph.

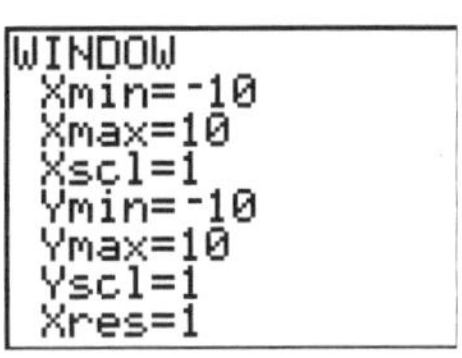

Figure 8-4: Window editor in Function mode.

   - **Xscl:** This is the distance between tick marks on the *x*-axis. (Go easy on the tick marks; using too many makes the axis look like a railroad track. Twenty or fewer tick marks makes for a nice looking axis.)

   - **Ymin and Ymax:** These are, respectively, the smallest and largest values of *y* that will be placed on the *y*-axis.

     If you have assigned values to **Xmin** and **Xmax** but don't know what values to assign to **Ymin** and **Ymax**, press [ZOOM][0] to invoke the **ZoomFit** command. This command uses the **Xmin** and

Xmax settings to determine the appropriate settings for **Ymin** and **Ymax**, and then automatically draws the graph. It does not change the **Yscl** setting. (You must return to the Window editor, if necessary, to adjust this setting yourself.)

- **Yscl:** This is the distance between tick marks on the *y*-axis. (As with the *x*-axis, too many tick marks make the axis look like a railroad track. Fifteen or fewer tick marks is a nice number for the *y*-axis.)
- **Xres:** This setting determines the resolution of the graph. It can be set to any of the integers 1 through 8. When **Xres** is set equal to 1, the calculator evaluates the function at each of the 94 pixels on the *x*-axis and graphs the result. If **Xres** is set equal to 8, the function is evaluated and graphed at every eighth pixel.

Xres is usually set equal to 1. If you're graphing a lot of functions, it may take the calculator a while to graph them at this resolution, but if you change **Xres** to a higher number, you may not get an accurate graph.

If it's taking a long time for the calculator to graph your functions, and this causes you to regret setting **Xres** equal to 1, press [ON] to terminate the graphing process. You can then go back to the Window editor and adjust the **Xres** setting to a higher number.

6. **Press [GRAPH] to graph the functions.**

Graphing Several Functions

If you're graphing several functions at once, it's not easy to determine which graph each function is responsible for. To help clear this up, the calculator allows you to identify the graphs of functions by setting a different graph style for each function. To do this, follow these steps:

1. **Press [Y=] to access the Y= editor.**
2. Use the [▶][◀][▲][▼] keys to place the cursor on the icon appearing at the far left of the definition of the function.
3. **Repeatedly press [ENTER] until you get the desired graph style.**

   You have seven styles to choose from: ⋱ (Line), ▜ (Thick Line), ▜ (shading above the curve), ▙ (shading below the curve), -0 (Path), 0 (Animate), and ⋱ (Dotted Line). Each time you press [ENTER], you get a different style.

   - **Line, Thick Line, and Dotted Line styles:** In Figure 8-2, $\mathbf{Y_1}$ is set to the default line style. $\mathbf{Y_2}$ and $\mathbf{Y_3}$ are set to Thick Line style; and $\mathbf{Y_6}$ is set to Dotted Line style. Figure 8-5 illustrates these styles.

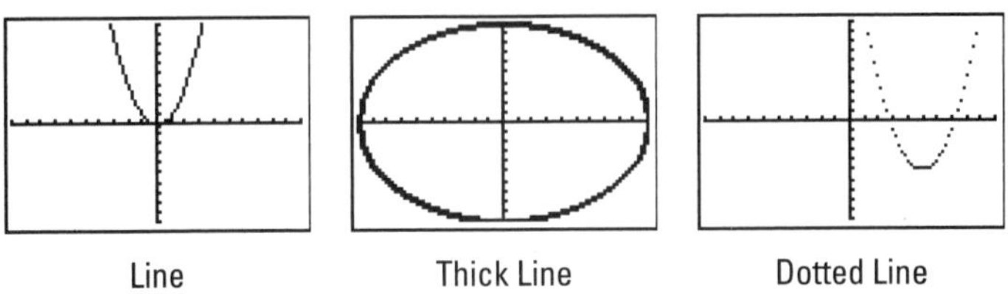

Figure 8-5: Line style, Thick Line style, and Dotted Line style.

   - **Shading above and below the curve styles:** In Figure 8-2, $\mathbf{Y_4}$ is set to the shading above the curve style and $\mathbf{Y_5}$ is set to the shading below the curve style. The calculator has four shading patterns: vertical lines, horizontal lines, negatively sloping diagonal lines, and positively sloping diagonal lines. You don't get to select the shading pattern. If you're graphing only one function in this style, the calculator uses the vertical line pattern. If you're graphing two functions, the first is graphed in the vertical line pattern and the second in the horizontal line pattern. If you graph three functions in this style, the third appears in the negatively sloping diagonal lines pattern, and so on (as illustrated in Figure 8-6).

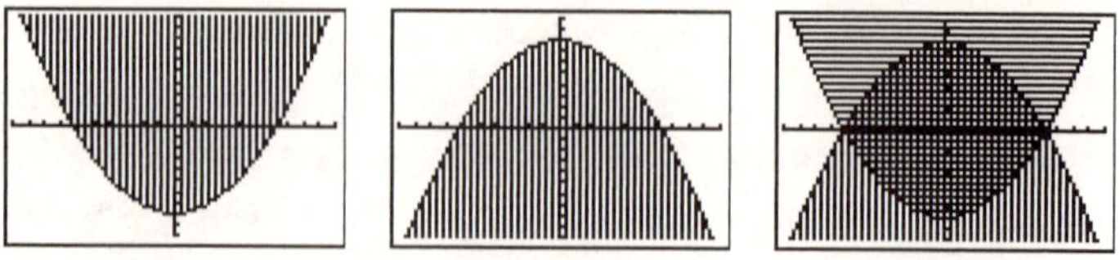

Above Below Both

Figure 8-6: Shading above the curve, shading below the curve, and two shaded functions.

- **Path and Animated styles:** The Path style, denoted by the -○ icon, uses a circle to indicate a point as it's being graphed (as illustrated in Figure 8-7). When the graph is complete, the circle disappears and leaves the graph in Line style.

  The animate style, denoted by the ○ icon, also uses a circle to indicate a point as it's being graphed, but when the graph is complete, no graph appears on the screen. For example, if this style is used, graphing $y = -x^2 + 9$ looks like a movie of the path of a ball thrown in the air.

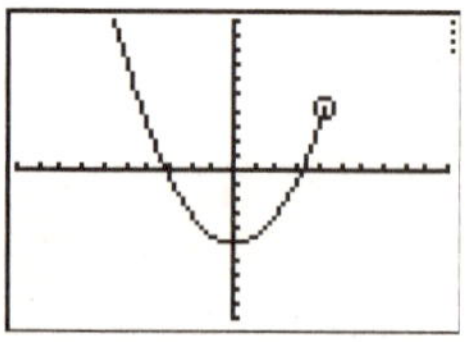

Figure 8-7: Path style.

TIP

If you don't want the calculator to graph a function in the Y= editor, remove the highlight from the equal sign in that function. To graph it at a later time, highlight the equal sign again. This is done in the Y= editor by using the [▶][◀][▲][▼] keys to place the cursor on the equal sign in the definition of the function and then pressing [ENTER] to toggle the equal sign between highlighted and not highlighted. In the example in Figure 8-2, the calculator won't graph functions $\mathbf{Y_2}$ through $\mathbf{Y_6}$.

Is Your Graph Accurate?

The calculator can do only what you tell it to do — which doesn't always produce an accurate graph. The three main causes of inaccurate graphs, and their solutions, are the following:

- **The graph is distorted by the size of the screen.**

  Because the calculator screen isn't square, circles don't look like circles unless the viewing window is properly set. How do you properly set the viewing window? No problem! Just graph the function as described earlier in this chapter, and then press [ZOOM][5] to invoke the **ZSquare** command. **ZSquare** readjusts the window settings for you and graphs the function again in a viewing window in which circles look like circles. Figure 8-8 illustrates this. (The circle being drawn in each of these figures is the circle defined by $\mathbf{Y}_2$ and $\mathbf{Y}_3$, as shown in Figure 8-2.)

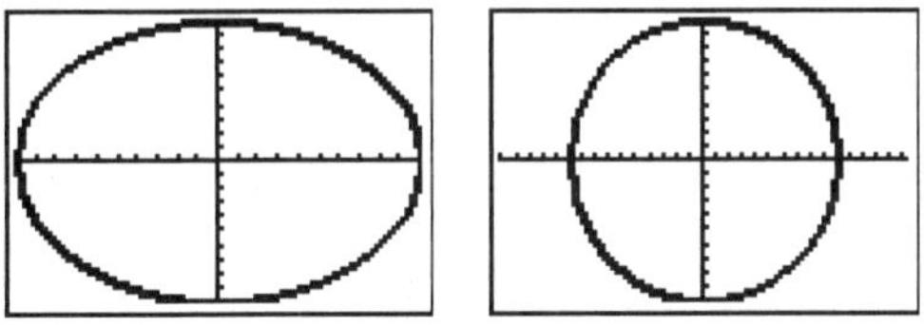

ZStandard ZSquare

Figure 8-8: A circle graphed using **ZStandard** and then using **ZSquare**.

- **The viewing window is too small or too big.**

  If you don't know what the graph should look like, then after graphing it you should zoom out to see more of the graph or zoom in to see a smaller portion of the graph. To do this, press [ZOOM][3] to zoom out, or press [ZOOM][2] to zoom in.

  Then use the [▶][◀][▲][▼] keys to move the cursor to the point from which you want to zoom out or in, and press [ENTER]. It's just like using a camera. The point you want to move the cursor to is the focal point.

  After zooming in or out, you may have to adjust the window settings, as described earlier in this chapter.

As an example, Figure 8-9 shows the progression of graphing $\mathbf{Y}_1$ in Figure 8-2. It was first graphed in the Standard viewing window. Then it was zoomed out from the point (0, 10). And finally, the window settings were adjusted to get a better picture of the graph.

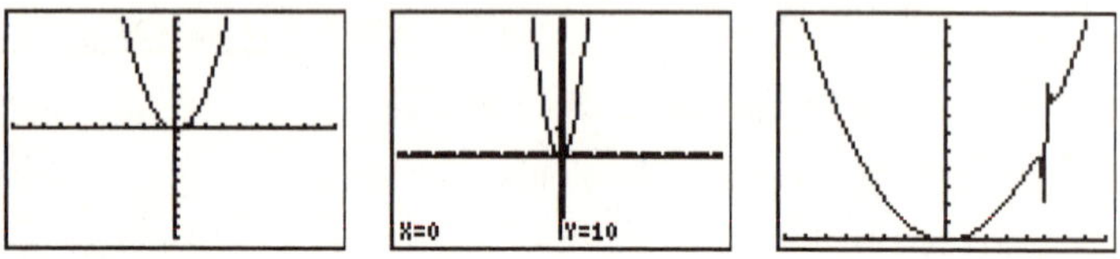

ZStandard | Zoom Out | Adjusted window

Figure 8-9: A graph using **ZStandard**, then **Zoom Out**, and then an adjusted viewing window.

- **Vertical asymptotes may not be recognizable.**

  In all graph styles except Animate and Dotted Line, the calculator graphs one point, and then the next point, and connects those two points with a line segment. This sometimes causes vertical asymptotes to appear on the graph. The last graph in Figure 8-9 illustrates an example of when a vertical asymptote is present. Don't mistake this almost-vertical line for a part of the graph. It's not; it's just a vertical asymptote.

  In a different viewing window, the vertical asymptote may not even appear. This happens when the calculator graphs one point, but the next point is undefined because the *x* value of that point is exactly at the location of the vertical asymptote. Figure 8-10 gives an example of re-graphing the last graph shown in Figure 8-9, using a viewing window in which the vertical asymptote does not appear.

  TIP: If you want to ensure that vertical asymptotes don't appear on your graph, graph the function in the Dotted Line style described in the previous section. For an explanation of how you can draw a vertical asymptote on a graph, see Chapter 12 in the book *TI-83 Plus Graphing Calculator For Dummies.*

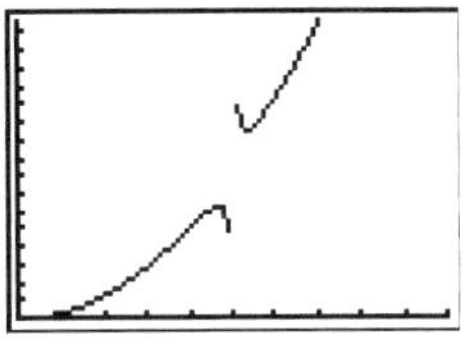

Figure 8-10: A viewing window in which no vertical asymptotes appear.

Piecewise-Defined Functions

When graphing a piecewise-defined function, some people cop out and simply graph each of the separate functions that appear in it. But this does not result in an accurate graph. To accurately graph the following piecewise-defined function, perform the following steps:

$$t = \begin{cases} y_1, x < a \\ y_2, a \leq x \leq b \\ y_3, x > b \end{cases}$$

1. **Enter the functions Y_1, Y_2, and Y_3 in the Y= editor.**

   Entering functions in the Y= editor is explained earlier in the chapter. Examples of three such functions appear in Figure 8-11.

2. **Turn off Y_1, Y_2, and Y_3 by removing the highlight from their equal signs.**

   This is done in the Y= editor by using the [▶][◀][▲][▼] keys to place the cursor on the equal sign in the definition of the function and then pressing [ENTER] to toggle the equal sign between highlighted and not highlighted. The calculator graphs a function only when its equal sign is highlighted. An example of this appears in Figure 8-11.

3. **Enter the piecewise-defined function in Y_4 as $(Y_1)(x < a) + (Y_2)(a \leq x)(x \leq b) + (Y_3)(x > b)$.**

The function must be entered as it appears above, complete with parentheses. **$(Y_1)(x < a)$** tells the calculator to graph the function Y_1 for $x < a$ and **$(Y_2)(a \le x)(x \le b)$** tells it to graph Y_2 for $a \le x \le b$.

- **To enter Y_1, Y_2, and Y_3:** Press [VARS][▸][1] and then press the number of the function you want to use in the definition.
- **To enter the inequalities:** Press [2nd][MATH] to access the Test menu, and then press the number of the inequality you want to use in the definition. For example, to enter the less-than (<) symbol, press [2nd][MATH][5].

An example of entering this function appears in Figure 8-11.

4. **Press [ENTER].**
5. **Graph the piecewise-defined function.**

   An earlier section of this chapter explains how to graph functions. An example graph of a piecewise-defined function appears in Figure 8-11.

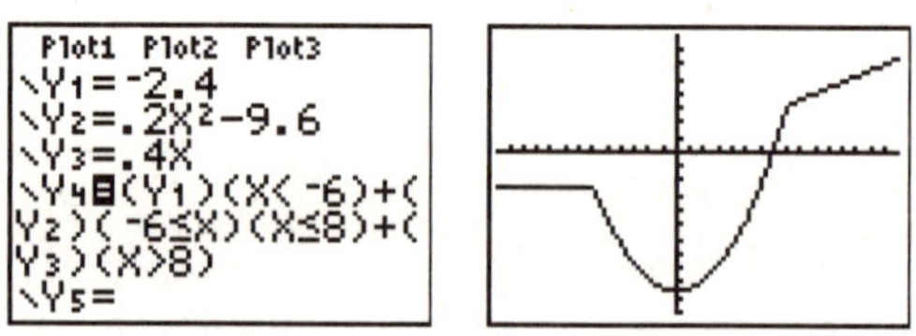

Definition Graph

Figure 8-11: Defining and graphing a piecewise-defined function.

TIP: If one or more of the functions in your piecewise-defined function is a trigonometric function, make sure the calculator is in Radian and not Degree mode. Otherwise, your piecewise-defined function may look like a step function instead of the graph you were expecting. The next section tells you how to change the mode and how to graph trigonometric functions.

Graphing Trig Functions

The calculator has built-in features especially designed for graphing trigonometric functions. They produce graphs that look like graphs you see in text books; and when you trace these graphs, the x-coordinate of the tracing point is always given as a fractional multiple of π. To use these features when graphing trigonometric functions, follow these steps:

1. **Put the calculator in Function and Radian mode.**

   Press [MODE]. In the third line, highlight **Radian**, and in the fourth highlight **Func**. (To highlight an item in the Mode menu, use the [▶][◀][▲][▼] keys to place the cursor on the item, and then press [ENTER].)

2. **Enter your trigonometric functions into the Y= editor.**

   Entering functions in the Y= editor is explained earlier in the chapter.

3. **Press [ZOOM][7] to graph the function.**

   [ZOOM][7] invokes the **ZTrig** command that graphs the function in a viewing window in which $-47\pi/24 \leq x \leq 47\pi/24$ and $-4 \leq y \leq 4$. It also sets the tick marks on the x-axis to multiples of $\pi/2$.

   When you trace a function graphed in a **ZTrig** window, the x-coordinate of the trace cursor will be a multiple of $\pi/24$, although the x-coordinate displayed at the bottom of the screen will be a decimal approximation of this value. (Tracing is explained in the next chapter.)

TIP

If you want to graph trigonometric functions in Degree mode, to get at least one period of a sine or cosine function, you must set **Xmin** to 0 and **Xmax** to 2π. This is why I say that it is a lot easier to graph trig functions in Radian mode. Setting the mode to degrees or radians is explained in Step 1.

Viewing the Function and Graph on the Same Screen

If you're planning to play around with the definition of a function you're graphing, it's quite handy to have both the Y= editor and the graph on the same screen. That way you can edit the definition of your function and see the effect your editing has on your graph. To do so, follow these steps:

1. **Put the calculator in Horizontal mode.**

   Press [MODE] and highlight **Horiz** in the last line of the menu, as illustrated in Figure 8-12. To highlight an item in the Mode menu, use the [▶][◀][▲][▼] keys to place the cursor on the item, and then press [ENTER].

2. **Press [Y=].**

   The Graph window appears at the top of the screen and the Y= editor at the bottom of the screen.

3. **Enter or edit a function in the Y= editor.**

   Entering functions in the Y= editor is explained earlier in this chapter. Editing expressions is explained in Chapter 1.

4. **Graph the function.**

   Graphing functions is explained earlier in this chapter.

To edit or enter a function, press [Y=]. To see the resulting graph, press [GRAPH].

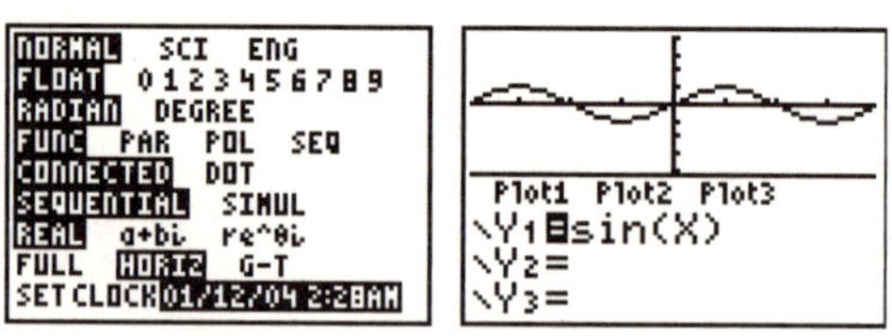

Mode Split screen

Figure 8-12: Function and graph on the same screen.

Saving and Recalling a Graph

If you want to save your graph as a Graph Database, when you recall the graph at a later time, the graph remains interactive. This means that you can, for example, trace the graph and resize the viewing window because a Graph Database also saves the Graph Mode, Window, Format, and Y= editor settings. It does not, however, save the split-screen settings (**Horiz** and **G-T**) entered in the last line of the Mode menu. This section explains how to save, delete, and recall a graph in a Graph Database.

To save a Graph Database, perform the following steps:

1. **Press [2nd][PRGM][▸][▸] to access the Draw Store menu.**
2. **Press [3] to store your graph as a Graph Database.**
3. **Enter an integer 0 through 9.**

   The calculator can store up to ten Graph Databases. If, for example, you enter the number **5**, your Graph Database is stored in the calculator as **GDB5**.

   If you save your Graph Database as **GDB5** without realizing that you had previously stored another Graph Database as **GDB5**, the calculator — without warning or asking your permission — erases the old **GDB5** and replaces it with the new **GDB5**. To see a list of the Graph Databases already stored in your calculator, press [2nd][+][2][9].

   If you already have ten Graph Databases stored in your calculator and don't want to sacrifice any of them, consider saving some of them on your PC.
4. **Press [ENTER].**

To delete a Graph Database from your calculator, perform the following steps:

1. **Press [2nd][+] to access the Memory menu.**
2. **Press [2] to access the Mem Mgt/Del menu.**
3. **Press [9] to access the GBD files stored in the calculator.**

4. **If necessary, repeatedly press ▼ to move the indicator to the GBD you want to delete.**

5. **Press DEL.**

   If there is more than one Graph Database stored in your calculator, you are asked whether or not you really want to delete this item. Press 2 if you want it deleted, or press 1 if you have changed your mind about deleting it.

6. **Press 2nd MODE to exit this menu and return to the home screen.**

To recall a saved Graph Database, perform the following steps:

1. **Press 2nd PRGM ▶ ▶ to access the Draw Store menu.**
2. **Press 4 to recall your Graph Database.**
3. **Enter the number of your stored Graph Database.**
4. **Press ENTER.**

   WARNING! When you recall a Graph Database, the Mode, Window, Format, and Y= editor settings in your calculator change to those saved in the Graph Database. If you don't want to lose the settings you have in the calculator, save them in another Graph Database before recalling your saved Graph Database. (Saving a Graph Database is described earlier in this section.)

Chapter 9

Exploring Functions

In This Chapter

- Using Zoom commands
- Tracing the graph of a function
- Constructing tables of functional values
- Creating and clearing user-defined tables
- Viewing graphs and tables on the same screen

The calculator has three very useful features that help you explore the graph of a function: zooming, tracing, and creating tables of functional values. Zooming allows you to quickly adjust the viewing window for the graph so that you can get a better idea of the nature of the graph. Tracing shows you the coordinates of the points that make up the graph. And creating a table — well, I'm sure you already know what that shows you. This chapter explains how to use each of these features.

Using Zoom Commands

After you've graphed your functions (as described in Chapter 8), you can use Zoom commands to adjust the view of your graph. Press [ZOOM] to see the ten Zoom commands that you can use. The following list explains the Zoom commands and how to use them:

- **Zoom commands that help you to initially graph or regraph your function in a preset viewing window:**
  - **ZStandard:** This command graphs your function in a preset viewing window where $-10 \leq x \leq 10$ and $-10 \leq y \leq 10$. You access it by pressing [ZOOM][6].

This is a nice Zoom command to use when you haven't the slightest idea what size viewing window to use for your function. After graphing the function using **ZStandard**, you can, if necessary, use the **Zoom In** and **Zoom Out** commands to get a better idea of the nature of the graph. Using **Zoom In** and **Zoom Out** are described later in this section.

• **ZDecimal:** This command graphs your function in a preset viewing window where $-4.7 \le x \le 4.7$ and $-3.1 \le y \le 3.1$. It is accessed by pressing [ZOOM][4].

When you trace a function graphed in a **ZDecimal** window, the *x*-coordinate of the trace cursor will be a multiple of 0.1. Tracing is explained in the next section.

• **ZTrig:** This command, which is most useful when graphing trigonometric functions, graphs your function in a preset viewing window where $-47\pi/24 \le x \le 47\pi/24$ and $-4 \le y \le 4$. It also sets the tick marks on the *x*-axis to multiples of $\pi/2$. You access **ZTrig** by pressing [ZOOM][7].

When you trace a function graphed in a **ZTrig** window, the *x*-coordinate of the trace cursor will be a multiple of $\pi/24$. Tracing is explained in the next section.

To use the zoom commands described above, enter your function into the calculator (as described in Chapter 8), press [ZOOM], and then press the key for the number of the command. The graph automatically appears.

- **Zoom commands that help you find an appropriate viewing window for the graph of your functions:**

• **ZoomFit:** This is my favorite Zoom command. If you know how you want to set the *x*-axis, **ZoomFit** automatically figures out the appropriate settings for the *y*-axis.

To use **ZoomFit**, press [WINDOW] and enter the values you want for **Xmin**, **Xmax**, and **Xscl**. Then press [ZOOM][0] to get **ZoomFit** to figure out the *y*-settings and graph your function. **ZoomFit** does not figure out an appropriate setting for **Yscl**, so you may want to go back to the Window editor and adjust this value. The Window editor is discussed in Chapter 8.

- **ZoomStat:** If you're graphing functions, this command is useless. But if you're graphing Stat Plots, this command finds the appropriate viewing window for your plots.

- **Zoom commands that readjust the viewing window of an already-graphed function:**

  - **ZSquare:** Because the calculator screen isn't perfectly square, graphed circles won't look like real circles unless the viewing window is properly set. **ZSquare** readjusts the existing Window settings for you and then regraphs the function in a viewing window in which circles look like circles.

    To use **ZSquare**, graph the function as described in Chapter 8, and then press [ZOOM][5]. The graph automatically appears.

  - **ZInteger:** This command is quite useful when you want the trace cursor to trace your functions using integer values of the *x*-coordinate, such as when graphing a function that defines a sequence. (Tracing is explained in the next section.) **ZInteger** readjusts the existing Window settings and regraphs the function in a viewing window in which the trace cursor displays integer values for the *x*-coordinate.

    To use **ZInteger**, graph the function as described in Chapter 8, and then press [ZOOM][8]. Use the [▶][◀][▲][▼] keys to move the cursor to the spot on the screen that will become the center of the new screen. Then press [ENTER]. The graph is redrawn centered at the cursor location.

- **Zoom commands that zoom in or zoom out from an already graphed function:**

  - **Zoom In and Zoom Out:** After the graph is drawn (as described in Chapter 8), these commands allow you to zoom in on a portion of the graph or to zoom out from the graph. They work very much like a zoom lens on a camera.

    Press [ZOOM][2] to zoom in or press [ZOOM][3] to zoom out. Then use the [▶][◀][▲][▼] keys to move the cursor to the spot on the screen from which you want to zoom in or zoom out. Then press [ENTER]. The graph is redrawn centered at the cursor location.

You can press [ENTER] again to zoom in closer or to zoom out one more time. Press [CLEAR] when you're finished zooming in or zooming out. You may have to adjust the window settings, as described earlier in Chapter 8.

- **ZBox:** This command allows you to define a new viewing window for a portion of your graph by enclosing it in a box, as illustrated in Figure 9-1. The box becomes the new viewing window.

  To construct the box, press [ZOOM][1] and use the [▶][◀][▲][▼] keys to move the cursor to the spot where you want one corner of the box to be located. Press [ENTER] to anchor that corner of the box. Then use the [▶][◀][▲][▼] keys to construct the rest of the box. When you press these keys, the calculator draws the sides of the box. Press [ENTER] when you're finished drawing the box. The graph is then redrawn in the window defined by your box (as shown in Figure 9-1).

  When you use **ZBox**, if you don't like the size of the box you get, you can use any of the [▶][◀][▲][▼] keys to resize the box. If you don't like the location of the corner you anchored, press [CLEAR] and start over.

  When you use **ZBox**, [ENTER] is pressed only two times. The first time you press it is to anchor a corner of the zoom box. The next time you press [ENTER] is when you're finished drawing the box, and you're ready to have the calculator redraw the graph.

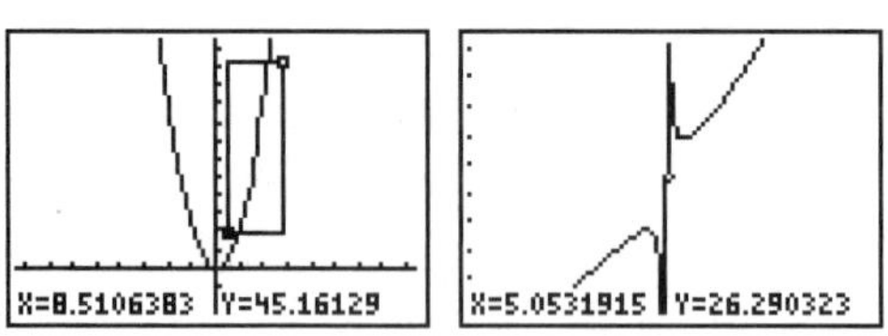

The box Redrawn graph

Figure 9-1: Constructing the Zoom Box and redrawing the graph.

TIP

Undoing a zoom

If you used a Zoom command to redraw a graph and then want to undo what that command did to the graph, follow these steps:

1. **Press [ZOOM][▶] to access the Zoom Memory menu.**
2. **Press [1] to select ZPrevious.**

The graph is redrawn as it appeared in the previous viewing window.

Tracing a Graph

After you have graphed your function, as described in the previous chapter, you can press [TRACE] and then use [▶] and [◀] to more closely investigate the function.

WARNING!

If you use only the [▶][◀][▲][▼] keys instead of [TRACE] to locate a point on a graph, all you will get is an *approximation* of the location of that point. You rarely get an actual point on the graph. So always use [TRACE] to identify points on a graph.

The following list describes what you see, or don't see, as you trace a graph:

- **The definition of the function:** The function you're tracing is displayed at the top of the screen, provided the calculator is in **ExprOn** format, as discussed in Chapter 8. If the Format menu is set to **ExprOff** and **CoordOn**, then the Y= editor number of the function appears at the top right of the screen. If the Format menu is set to **ExprOff** and **CoordOff**, then tracing the graph is useless because all you see is a cursor moving on the graph. The calculator won't tell you which function you're tracing, nor will it tell you the coordinates of the cursor location. (The Format menu and Y= editor are described in Chapter 8.)

  If you've graphed more than one function and you would like to trace a different function, press [▲]. Each time you press this key, the cursor jumps to another function. Eventually it jumps back to the original function.

- **The values of x and y:** At the bottom of the screen, you see the values of the x- and y-coordinates that define the cursor location, provided the calculator is in **CoordOn** format (as discussed in Chapter 8). In the **PolarGC** format, the coordinates of this point display in polar form.

  When you press [TRACE], the cursor is placed on the graph at the point having an x-coordinate that is approximately midway between **Xmin** and **Xmax**. If the y-coordinate of the cursor location isn't between **Ymin** and **Ymax**, then the cursor does not appear on the screen. The sidebar, "Panning in Function mode," tells you how to correct this situation.

  Each time you press [▶], the cursor moves right to the next plotted point on the graph, and the coordinates of that point are displayed at the bottom of the screen. If you press [◀], the cursor moves left to the previously plotted point. And if you press [▲] to trace a different function, the tracing of that function starts at the point on the graph that has the x-coordinate displayed on-screen before you pressed this key.

Press [CLEAR] to terminate tracing the graph. This also removes the name of the function and the coordinates of the cursor from the screen.

TIP: When you're using TRACE, if you want to start tracing your function at a specific value of the independent variable x, just key in that value and press [ENTER] when you're finished. (The value you assign to x must be between **Xmin** and **Xmax**; if it's not, you get an error message.) After you press [ENTER], the trace cursor moves to the point on the graph having the x-coordinate you just entered. If that point isn't on the portion of the graph appearing on the screen, the sidebar, "Panning in Function mode," tells you how to get the cursor and the graph in the same viewing window.

TIP: If the name of the function and the values of x and y are interfering with your view of the graph when you use TRACE, increase the height of the screen by pressing [WINDOW], and then decrease the value of **Ymin** and increase the value of **Ymax**.

TIP

Panning in Function mode

When you're tracing a function and the cursor hits the top or bottom of the screen, you will still see the coordinates of the cursor location displayed at the bottom of the screen but you won't see the cursor itself on the screen because the viewing window is too small. Press [ENTER] to get the calculator to adjust the viewing window to a viewing window that is centered about the cursor location. If the function you were tracing isn't displayed at the top of the screen, repeatedly press [▲] until it is. The trace cursor then appears in the middle of the screen and you can use [▶] and [◀] to continue tracing the graph.

When you're tracing a function and the cursor hits the left or right side of the screen, the calculator automatically pans left or right. It also appropriately adjusts the values assigned to **Xmin** and **Xmax** in the Window editor, but it does not change the values of **Ymin** and **Ymax**. So you may not see the cursor on the screen. If this happens, follow the directions in the previous paragraph to see both the function and the cursor on the screen.

Displaying Functions in a Table

After you've entered the functions in the Y= editor (as described in Chapter 8), you can have the calculator create a table of functional values. To create a table, perform the following steps:

1. **Highlight the equal sign of those functions in the Y= editor that you want to appear in the table.**

   Only those functions in the Y= editor that are defined with a highlighted equal sign appear in the table. To highlight or remove the highlight from an equal sign, press [Y=], use the [▶][◀][▲][▼] keys to place the cursor on the equal sign in the definition of the function, and then press [ENTER] to toggle the equal sign between highlighted and not highlighted.

2. **Press [2nd][WINDOW] to access the Table Setup editor (shown in Figure 9-2).**

Figure 9-2: The Table Setup editor.

3. **Enter a number in TblStart, and then press [ENTER].**

   TblStart is the first value of the independent variable *x* to appear in the table. In Figure 9-2, **TblStart** is assigned the value 5.

   To enter the number you have chosen for **TblStart**, place the cursor on the number appearing after the equal sign, press the number keys to enter your new number, and then press [ENTER].

4. **Enter a number in ΔTbl, and then press [ENTER].**

   ΔTbl gives the increment for the independent variable *x*. In Figure 9-2, **ΔTbl** is assigned the value –1.

   To enter the number you have chosen for **ΔTbl**, place the cursor on the number appearing after the equal sign, press the number keys to enter your new number, and then press [ENTER].

5. **Set the mode for Indpnt and Depend.**

   To change the mode of either **Indpnt** or **Depend**, use the [▶][◀][▲][▼] keys to place the cursor on the desired mode, either **Auto** or **Ask**, and then press [ENTER].

   To have the calculator automatically generate the table for you, put both **Indpnt** and **Depend** in **Auto** mode. The table in Figure 9-3 was constructed in this fashion.

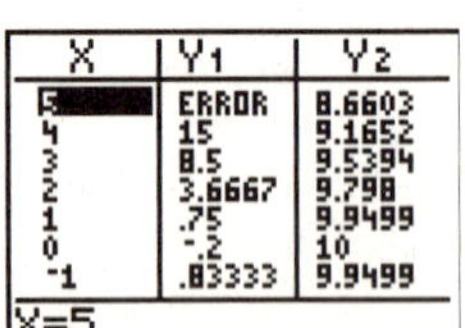

Figure 9-3: A table automatically generated by the calculator.

If you want to create a user-defined table in which you specify which values of the independent variable *x* appear in the table — and then have the calculator figure out the corresponding values of the functions — put **Indpnt** in **Ask** mode and **Depend** in **Auto** mode. (How you construct the table is explained in Step 6.) The table in Figure 9-4 was constructed in this fashion.

TIP

For a user-defined table, you don't have to assign values to **TblStart** and **ΔTbl** in the Table Setup editor.

The other combinations of mode settings for **Indpnt** and **Depend** are not all that useful, unless you want to play a quick round of "Guess the *y*-coordinate."

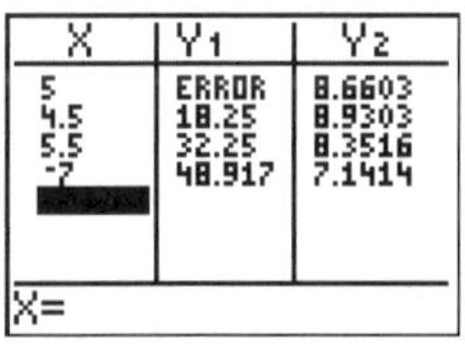

X	Y1	Y2
5	ERROR	8.6603
4.5	18.25	8.9303
5.5	32.25	8.3516
-7	48.917	7.1414

X=

Figure 9-4: A user-generated table.

6. **Press [2nd][GRAPH] to display the table.**

   When you display the table, what you see on the screen depends on the modes you set for **Indpnt** and **Depend** in Step 5. And what you can do with the table also depends on these modes. Here's what you see and what you can do:

   - **An automatically generated table:**

     If **Indpnt** and **Depend** are both in **Auto** mode, then when you press [2nd][GRAPH], the table is automatically generated. To display rows in the table beyond the last row on the screen, repeatedly press [▼] until they appear. You can repeatedly press [▲] to display rows above the first row on the screen.

     TIP

     Each time the calculator redisplays a table with a different set of rows, it also automatically resets **TblStart** to the value of *x* that appears in the first row of the newly displayed table. To return the table to its original state, press [2nd][WINDOW] to access the Table Setup editor, and then change the value that the calculator assigned to **TblStart**.

If you're constructing a table for more than two functions, only the first two functions appear on the screen. To see the other functions, repeatedly press [▶] until they appear. This causes one or more of the initial functions to disappear. To see them again, repeatedly press [◀] until they appear.

- **A user-generated table:**

  If you put **Indpnt** in **Ask** mode and **Depend** in **Auto** mode so that you can generate your own table, then when you display the table, it should be empty. If it's not empty, clear the table (as described in the next section).

  In an empty table, key in the first value of the independent variable *x* that you want to appear in the table, and then press [ENTER]. The corresponding *y*-values of the functions in the table automatically appear. Key in the next value of *x* you want in the table and press [ENTER], and so on. The values of *x* that you place in the first column of the table don't have to be in any specific order, nor do they have to be between the **Xmin** and **Xmax** settings in the Window editor.

TIP: The word ERROR appearing in a table doesn't indicate that the creator of the table has done something wrong. It indicates that either the function is undefined or the corresponding value of *x* is not a real-valued number. This is illustrated in Figures 9-3 and 9-4.

TIP: While displaying the table of functional values, you can edit the definition of a function without going back to the Y= editor. To do this, use the [▶][◀][▲][▼] keys to place the cursor on the column heading for that function and then press [ENTER]. Edit the definition of the function (editing expressions is explained in Chapter 1) and press [ENTER] when you're finished. The calculator automatically updates the table and the definition of the function in the Y= editor.

Clearing a Table

Not all tables are created alike. An automatically generated table, for example, cannot be cleared. To change the contents of such a table, you have to change the values assigned to **TblStart** and **ΔTbl** in the Table Setup editor. After you have created a user-defined table, however, you can perform the following steps to clear its contents:

1. **Press [2nd][WINDOW] to access the Table Setup editor, and then set Indpnt to Auto.**
2. **Press [2nd][GRAPH] to display an automatically generated table.**
3. **Press [2nd][WINDOW] and set Indpnt back to Ask.**
4. **Press [2nd][GRAPH] to display an empty table.**

Viewing the Table and the Graph on the Same Screen

After you have graphed your functions and created a table of functional values, you can view the graph and the table on the same screen. To do so, follow these steps:

1. **Press [MODE].**
2. **Put the calculator in G-T screen mode.**

   To do so, use the [▶][◀][▲][▼] keys to place the cursor on **G-T** in the lower-right corner of the Mode menu, and then press [ENTER] to highlight it. This is illustrated in Figure 9-5.

Figure 9-5: Setting the mode for viewing a graph and a table.

3. **Press GRAPH.**

   After you press GRAPH, the graph and the table appear on the same screen (as shown in Figure 9-6).

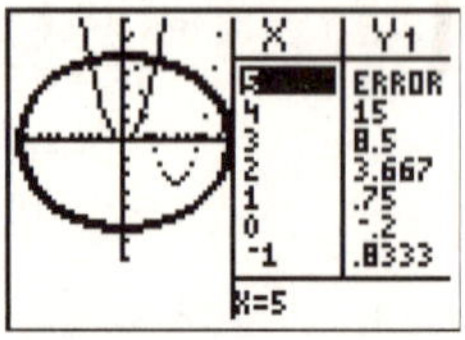

Figure 9-6: A Graph-Table split screen.

If you press any key used in graphing functions, such as ZOOM or TRACE, the cursor becomes active on the graph side of the screen. To return the cursor to the table, press 2nd GRAPH.

If you press TRACE and then use the ▶ ◀ ▲ ▼ keys to trace the graph, the value of the independent variable *x* corresponding to the cursor location on the graph is highlighted in the table and the column for the function you're tracing appears next to it. If necessary, the calculator updates the table so you can see that row in the table (as in Figure 9-7).

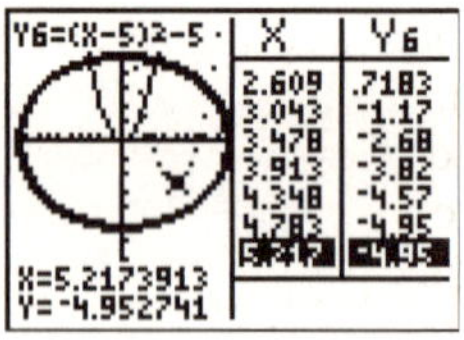

Figure 9-7: Using TRACE in Graph-Table mode.

To view the graph or the table in full screen mode, you can use these steps:

1. **Press MODE.**
2. **Put the calculator in Full screen mode.**

   To do so, use the ▶ ◀ ▲ ▼ keys to place the cursor on **Full** in the bottom left hand corner of the Mode menu and press ENTER to highlight it.
3. **Press GRAPH to see the graph, or press 2nd GRAPH to see the table.**

Chapter 10

Evaluating Functions

In This Chapter

- Finding the value of a function
- Finding the zeros (*x*-intercepts) of a function
- Finding the maximum and minimum values of a function
- Finding the point of intersection of two functions
- Finding the slope of a tangent to the graph of a function
- Finding the value of the definite integral of a function

After graphing a function (as described in Chapter 8), you can use the options on the Calculate menu to find the value of the function at a specified value of x, to find the zeros (x-intercepts) of the function, and to find the maximum and minimum values of the function. You can even find the derivative of the function at a specified value of x, or you can evaluate a definite integral of the function. This, in turn, enables you to find the slope of the tangent to the graph of the function at a specified value of x or to find the area between the graph and the x-axis. Moreover, if you have graphed two functions, there is an option on the Calculate menu that finds the coordinates of these two functions' points of intersection.

The rest of this chapter tells you how to use the Calculate menu to find these values. But be warned: The calculator is not perfect. In most cases, using the options on the Calculate menu yields only an approximation of the true value (albeit a very *good* approximation).

Finding the Value of a Function

When you trace the graph of a function, the trace cursor doesn't hit every point on the graph. So tracing is not a reliable way of finding the value of a function at a specified value of the independent variable *x*. The **CALC** menu, however, contains a command that will evaluate a function at any specified *x*-value. To access and use this command, perform the following steps:

1. **Graph the functions in a viewing window that contains the specified value of *x*.**

   Graphing functions and setting the viewing window are explained in Chapter 8. To get a viewing window containing the specified value of *x*, that value must be between **Xmin** and **Xmax**.

2. **Set the Format menu to ExprOn and CoordOn.**

   Setting the Format menu is explained in Chapter 8.

3. **Press [2nd][TRACE] to access the Calculate menu.**

4. **Press [1] to select the value option.**

5. **Enter the specified value of *x*.**

   REMEMBER: When using the **value** command to evaluate a function at a specified value of *x*, that value must be an *x*-value that appears on the *x*-axis of the displayed graph — that is, it must be between **Xmin** and **Xmax**. If it isn't, you get an error message.

   Use the keypad to enter the value of *x* (as illustrated in the first graph in Figure 10-1). If you make a mistake when entering your number, press [CLEAR] and re-enter the number.

6. **Press [ENTER].**

   After you press [ENTER], the first highlighted function in the Y= editor appears at the top of the screen, the cursor appears on the graph of that function at the specified value of *x*, and the coordinates of the cursor appear at the bottom of the screen. This is illustrated in the second graph in Figure 10-1.

7. **Repeatedly press [▲] to see the value of the other graphed functions at your specified value of *x*.**

   Each time you press [▲], the name of the function being evaluated appears at the top of the screen and the coordinates of the cursor location appear at the bottom of the screen. This is illustrated in the third graph in Figure 10-1.

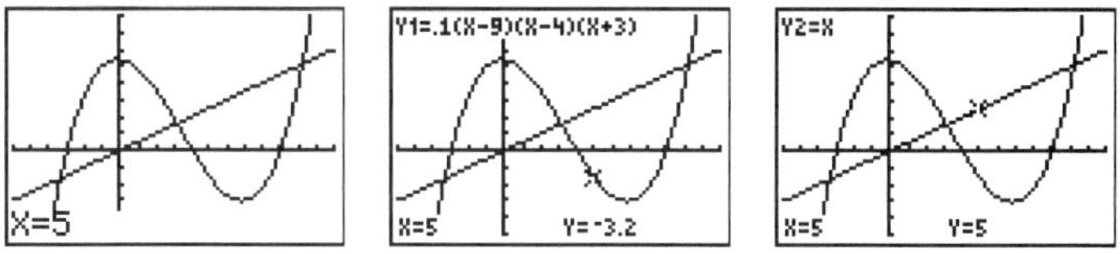

Enter *x* | Press [ENTER] | Press [▲]

Figure 10-1: Steps in evaluating two functions at a specified value of *x*.

After using the **value** command to evaluate your functions at one value of *x*, you can evaluate your functions at another value of *x* by keying in the new value and pressing [ENTER]. Pressing any function key (such as [ENTER] or [TRACE]) *after* evaluating a function deactivates the **value** command.

TIP: If you plan to evaluate functions at several specified values of *x*, consider constructing a user-defined table of functional values (as explained in Chapter 9).

Finding the Zeros of a Function

The *zeros* of the function $y = f(x)$ are the solutions to the equation $f(x) = 0$. Because $y = 0$ at these solutions, these zeros (solutions) are really just the *x*-coordinates of the *x*-intercepts of the graph of $y = f(x)$. (An *x*-intercept is a point where the graph crosses or touches the *x*-axis.) To find a zero of a function, perform the following steps:

1. **Graph the function in a viewing window that contains the zeros of the function.**

   Graphing a function and finding an appropriate viewing window are explained in Chapter 8. To get a viewing window containing a zero of the function, that zero must be between **Xmin** and **Xmax** and the *x*-intercept at that zero must be visible on the graph.

2. **Set the Format menu to ExprOn and CoordOn.**

   Setting the Format menu is explained in Chapter 8.

3. **Press [2nd][TRACE] to access the Calculate menu.**

4. **Press [2] to select the zero option.**

5. **If necessary, repeatedly press [▲] until the appropriate function appears at the top of the screen.**

6. **Set the Left Bound for the zero you desire to find.**

   To do so, use the [◄] and [►] keys to place the cursor on the graph a little to the left of the zero, and then press [ENTER]. A Left Bound indicator appears at the top of the screen (as illustrated by the triangle in the first graph of Figure 10-2).

7. **Set the Right Bound for the zero.**

   To do so, use the [◄] and [►] keys to place the cursor on the graph a little to the right of the zero, and then press [ENTER]. A Right Bound indicator appears at the top of the screen, as shown by the right triangle in the second graph of Figure 10-2.

8. **Tell the calculator where you guess the zero is located.**

   This guess is necessary because the calculator uses a numerical routine for finding a zero. The routine is an iterative process that requires a seed (guess) to get it started. The closer the seed is to the zero, the faster the routine finds the zero. To do this, use the [◄] and [►] keys to place the cursor on the graph as close to the zero as possible, and then press [ENTER]. The value of the zero appears at the bottom of the screen, as shown in the third graph of Figure 10-2.

TIP: The calculator uses scientific notation to denote really large or small numbers. For example, –0.00000001 is displayed on the calculator as –1E–8, and 0.000000005 is displayed as 5E–8.

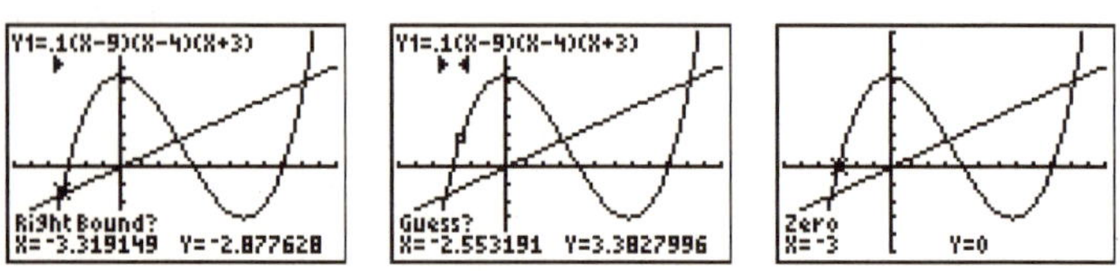

Left bound Right bound Guess

Figure 10-2: Steps in finding the zero of a function.

Finding Min & Max

Finding the maximum or minimum point on a graph has many useful applications. For example, the maximum point on the graph of a profit function not only tells you the maximum profit (the *y*-coordinate), it also tells you how many items (the *x*-coordinate) the company must manufacture to achieve this profit. To find the minimum or maximum value of a function, perform the following steps:

1. **Graph the function in a viewing window that contains the minimum and/or maximum values of the function.**

   Graphing a function and finding an appropriate viewing window are explained in Chapter 8.

2. **Set the Format menu to ExprOn and CoordOn.**

   Setting the Format menu is explained in Chapter 8.

3. **Press [2nd][TRACE] to access the Calculate menu.**

4. **Press [3] to find the minimum, or press [4] to find the maximum.**

5. **If necessary, repeatedly press [▲] until the appropriate function appears at the top of the screen.**

6. **Set the Left Bound of the minimum or maximum point.**

   To do so, use the [◀] and [▶] keys to place the cursor on the graph a little to the left of the location of the minimum or maximum point, and then press [ENTER]. A *left bound indicator* (the triangle shown in the first graph of Figure 10-3) appears at the top of the screen.

7. **Set the Right Bound for the zero.**

   To do so, use the [◀] and [▶] keys to place the cursor on the graph a little to the right of the location of the minimum or maximum point, and then press [ENTER]. A *right bound indicator* (the rightmost triangle in the second graph of Figure 10-3) appears at the top of the screen.

8. **Tell the calculator where you guess the min or max is located.**

   To do so, use the [◀] and [▶] keys to place the cursor on the graph as close to the location of the minimum or maximum point as possible, and then press ENTER. The coordinates of the minimum or maximum point appear at the bottom of the screen (as in the third graph of Figure 10-3).

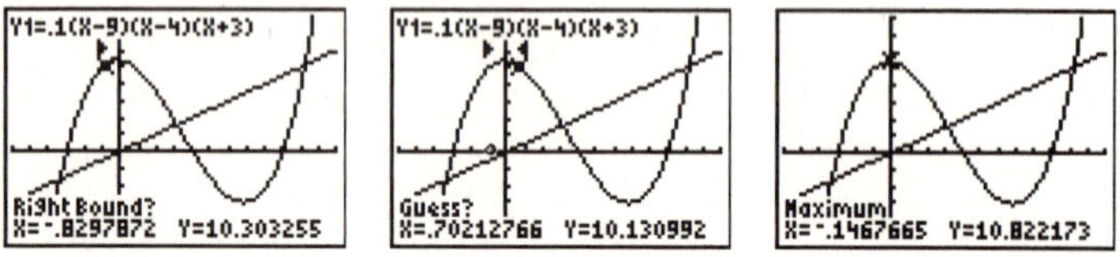

Left bound Right bound Guess

Figure 10-3: Steps in finding the maximum value of a function.

Finding Points of Intersection

Using the ▶◀▲▼ keys to locate the point of intersection of two graphs gives you an *approximation* of that point, but this method rarely gives you the actual point of intersection. To accurately find the coordinates of the point where two functions intersect, perform the following steps:

1. **Graph the functions in a viewing window that contains the point of intersection of the functions.**

   Graphing a function and finding an appropriate viewing window are explained in Chapter 8.

2. **Set the Format menu to ExprOn and CoordOn.**

   Setting the Format menu is explained in Chapter 8.

3. **Press 2nd TRACE to access the Calculate menu.**

4. **Press 5 to select the intersect option.**

5. **Select the first function.**

   If the name of one of the intersecting functions does not appear at the top of the screen, repeatedly press ▲ until it does. This is illustrated in the first graph of Figure 10-4. When the cursor is on one of the intersecting functions, press ENTER to select it.

6. **Select the second function.**

   If the calculator does not automatically display the name of the second intersecting function at the top of the screen, repeatedly press [▲] until it does. This is illustrated in the second graph of Figure 10-4. When the cursor is on the second intersecting function, press [ENTER] to select it.

7. **Use the [◄] and [►] to move the cursor as close to the point of intersection as possible.**

   This is illustrated in the third graph in Figure 10-4.

8. **Press [ENTER] to display the coordinates of the point of intersection.**

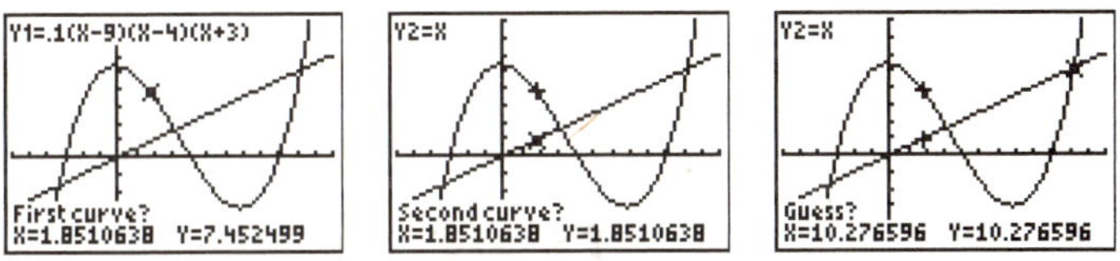

First function — Second function — Guess

Figure 10-4: Steps in finding a point of intersection.

Finding the Slope of a Curve

The calculator is not equipped to find the derivative of a function. For example, it cannot tell you that the derivative of x^2 is $2x$. But the calculator is equipped with a numerical routine that evaluates the derivative at a specified value of x. This numerical value of the derivative is the slope of the tangent to the graph of the function at the specified x-value. It is also called the slope of the curve. To find the slope (derivative) of a function at a specified value of x, perform the following steps:

1. **Graph the function in a viewing window that contains the specified value of x.**

   Graphing a function and setting the viewing window are explained in Chapter 8. To get a viewing window containing the specified value of x, that value must be between **Xmin** and **Xmax**.

2. **Set the Format menu to ExprOn and CoordOn.**

   Setting the Format menu is explained in Chapter 8.

3. **Press [2nd][TRACE] to access the Calculate menu.**
4. **Press [6] to select the dy/dx option.**
5. **If necessary, repeatedly press [▲] until the appropriate function appears at the top of the screen.**

   This is illustrated in the first graph in Figure 10-5.
6. **Enter the specified value of *x*.**

   To do so, use the keypad to enter the value of *x*. As you use the keypad, **X=** appears, replacing the coordinates of the cursor location appearing at the bottom of the screen in the previous step. The number you key in appears after **X=**. This is illustrated in the second graph in Figure 10-5. If you make a mistake when entering your number, press [CLEAR] and re-enter the number.

   TIP: If you are interested only in finding the slope of the function in a general area of the function instead of at a specific value of *x*, instead of entering a value of *x*, just use the [◄] and [►] to move the cursor to the desired location on the graph of the function.
7. **Press [ENTER].**

   After pressing [ENTER], the slope (derivative) is displayed at the bottom of the screen. This is illustrated in the third graph in Figure 10-5.

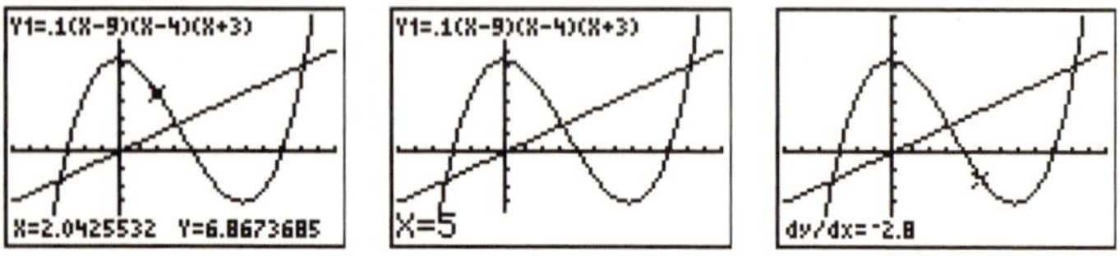

Select function   Enter *x*   Press [ENTER]

Figure 10-5: Steps in finding the slope at a specified value of *x*.

Evaluating a Definite Integral

If $f(x)$ is positive for $a \le x \le b$, then the definite integral $\int_a^b f(x)\,dx$ also gives the area between the curve and the *x*-axis for $a \le x \le b$. To evaluate the definite integral, perform the following steps:

1. **Graph the function *f*(*x*) in a viewing window that contains the lower limit *a* and the upper limit *b*.**

   Graphing a function and setting the viewing window are explained in Chapter 8. To get a viewing window containing *a* and *b*, these values must be between **Xmin** and **Xmax**.

2. **Set the Format menu to ExprOn and CoordOn.**

   Setting the Format menu is explained in Chapter 8.

3. **Press [2nd][TRACE] to access the Calculate menu.**

4. **Press [7] to select the $\int f(x)dx$ option.**

5. **If necessary, repeatedly press [▲] until the appropriate function appears at the top of the screen.**

   This process is illustrated in the first graph in Figure 10-6.

6. **Enter the value of the lower limit *a*.**

   To do so, use the keypad to enter the value of the lower limit *a*. As you use the keypad, **X=** appears, replacing the coordinates of the cursor location appearing at the bottom of the screen in the previous step. The number you key in appears after **X=**. This is illustrated in the second graph in Figure 10-6. If you make a mistake when entering your number, press [CLEAR] and re-enter the number.

7. **Press [ENTER].**

   After pressing [ENTER], a left bound indicator appears at the top of the screen.

8. **Enter the value of the upper limit *b* and press [ENTER].**

   After pressing [ENTER], the value of the definite integral appears at the bottom of the screen and the area between the curve and the *x*-axis, for $a \le x \le b$, will be shaded. This is illustrated in the third graph in Figure 10-6.

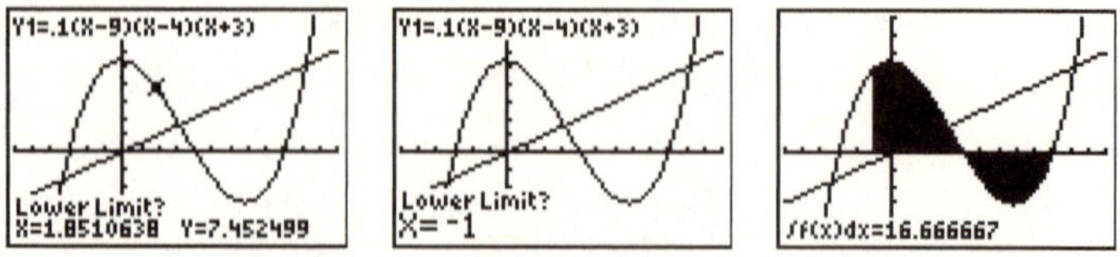

Select function Enter limits Press [ENTER]

Figure 10-6: Steps in evaluating a definite integral.

The shading of the graph produced by using the $\int f(x)dx$ option on the Calculate menu does not automatically vanish when you use another Calculate option. To erase the shading, press [2nd][PRGM][1] to invoke the **ClrDraw** command on the Draw menu. The graph is then redrawn without the shading.

Chapter 11

Probability

In This Chapter

- Evaluating permutations and combinations
- Generating random numbers

Do you need to calculate the number of ways you can arrange six people at a table or the number of ways you can select four people from a group of six people? Or do you just need an unbiased way of selecting people at random? If so, this is the chapter for you.

Permutations and Combinations

A *permutation,* denoted by **nPr**, answers the question: "From a set of **n** different items, how many ways can you select *and* order (arrange) **r** of these items?" A *combination,* denoted by **nCr**, answers the question: "From a set of **n** different items, how many ways can you select (independent of order) **r** of these items?" To evaluate a permutation or combination, follow these steps:

1. **On the Home screen, enter n, the total number of items in the set.**

   If you're not already on the Home screen, press [2nd][MODE] to exit (quit) the current screen and enter the Home screen.

2. **Press [MATH][▸][▸][▸] to access the Math Probability menu.**

3. **Press [2] to evaluate a permutation or press [3] to evaluate a combination.**

4. **Enter r, the number of items selected from the set, and press [ENTER] to display the result.**

   Figure 11-1 illustrates this procedure. Notice that the last two lines in the figure tell you that it's impossible to select seven items from a set of only five items.

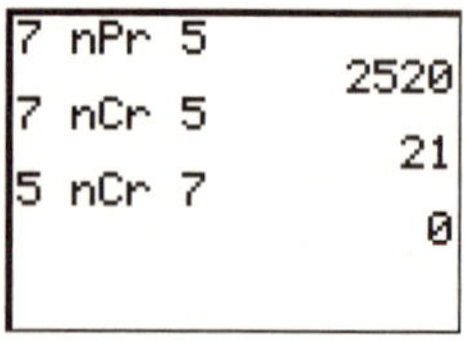

Figure 11-1: Evaluating permutations and combinations.

Generating Random Numbers

When generating random numbers, you usually want to generate numbers that are integers contained in a specified range, or decimal numbers that are strictly between 0 and 1.

Generating random integers

To generate random integers that fall between the integers **a** and **b**:

1. **Press [MATH][▸][▸][▸][5] to select the randInt command from the Math Probability menu.**
2. **Enter the value of the lower limit a, press [,], and enter the upper limit b.**
3. **Press [ENTER] to generate the first random integer. Repeatedly press [ENTER] to generate more random integers.**

   This is illustrated in the first picture in Figure 11-2.

Pressing any key except [ENTER] stops the calculator from generating random integers.

Generating random decimals

To generate random decimal numbers that are strictly between 0 and 1, press [MATH][▸][▸][▸][1] to select the **rand** command from the Math Probability menu. Then repeatedly press [ENTER] to generate the random numbers. The second picture in Figure 11-2 illustrates this process.

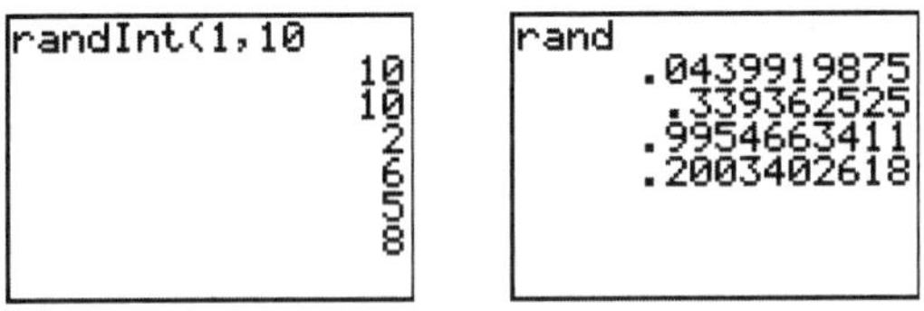

Integers Decimals

Figure 11-2: Generating random numbers.

Chapter 12

Simulating Probabilities

In This Chapter

- Tossing coins
- Rolling dice
- Picking marbles from a pot
- Spinning a spinner
- Drawing cards from a deck

With the Probability Simulation application that comes preloaded on the TI-84 Plus family of calculators, you can simulate tossing coins, rolling dice, picking marbles from a pot, spinning a spinner, and drawing cards from a deck. It even allows you to rig the simulation with weighted coins and unfair dice.

As the application performs simulations, it collects the data associated with these simulations and displays it in a graph or table. It can even save the data for you so that you can perform a statistical analysis of the data at a later time. In this chapter, I tell you how to get the Probability Simulation application to do all these neat things.

Starting and Quitting Probability Simulation

To start the Probability Simulation application, press [APPS][ALPHA][8], use [▼] to move the cursor to the **Prob Sim** application, and press [ENTER] to select the application and display the logo screen, as illustrated in the first screen in

Figure 12-1. Press any key to start the Probability Simulation application and display the Main menu screen, as illustrated in the second picture in this figure.

Logo screen Main menu

Figure 12-1: Screens you see when you start Probability Simulation.

You quit the Probability Simulation application by using the **QUIT** command at the bottom of the Main menu, as illustrated in the second picture in Figure 12-1. If you are not in the Main menu, repeatedly press [Y=] until this menu appears. Then press [GRAPH] to exit the application. When asked if you really want to quit, press [Y=] to select the **YES** option.

Using the Probability Simulation Application

Before you toss coins, roll dice, pick marbles, spin spinners, or draw cards, you need to know how to access the commands at the bottom of the screen, how to seed the random number generator, and what the **ESC** command does. These things are explained in this section; tossing coins, rolling dice, picking marbles, spinning spinners, drawing cards, and generating random numbers are explained later in this chapter.

Executing the commands at the bottom of the screen

At the bottom of each screen, you see a line of five boxes containing from two to five commands, as illustrated in the second picture in Figure 12-1 where the boxes contain four commands (**OK**, **OPTN**, **ABOUT**, and **QUIT**) with no command in the second box. Under these boxes are five calculator keys: [Y=],

[WINDOW], [ZOOM], [TRACE], and [GRAPH]. You execute the command by pressing the calculator key that is under the box. For example, in the second picture in Figure 12-1, the **OPTN** command is executed by pressing [ZOOM].

Seeding the random number generator

The calculator doesn't really generate random numbers; it uses a formula to generate a sequence of numbers that appear to be random. Each number in this sequence is dependent on the number that precedes it. Thus the whole sequence is determined by the first number in the sequence. This number is called the seed; assigning a value to the seed is called seeding the random number generator. If you seed two different TI-84s with the same number, the random number generator will produce the same sequence of numbers on each calculator. Now that isn't very random, is it?

The outcome of each simulation performed by the Probability Simulation application, whether it be tossing a coin or spinning a spinner, is determined by a random number generated by Probability Simulation's random number generator. Usually, you don't need to seed the random number generator to ensure that you get a "random" outcome. You can just accept whatever number the calculator uses as a seed because it uses the last random number it generated as the seed.

However, if you want to ensure that a simulation is replicated on two or more TI-84s, set the seed to the same number on each calculator. Or if you want to ensure that a simulation is not replicated, set the seed to a random number of your choice.

To seed the random number generator, follow these steps:

1. **Go to Probability Simulation's Main menu.**

   The Main menu, pictured in the second screen in Figure 12-1, is the first screen you see when you start the Probability Simulation application. If you are not in the Main menu, repeatedly press [Y=] until this menu appears.

2. **Press [ZOOM] to execute the OPTN command.**

   A blinking cursor appears, waiting for you to enter a number.

3. **Use the keypad to enter any positive integer.**

4. **Press [GRAPH] to select the OK option.**

   The random number generator is seeded by the number you entered, and you are returned to the Main menu.

Going back to the previous screen using the ESC command

The **ESC** command sends you back to the previous screen. If it is available, it is always the first command in the command line at the bottom of the screen. So it is always executed by pressing [Y=].

Understanding the Settings Editor

The Probability Simulation application contains four simulators (Tossing Coins, Rolling Dice, Picking Marbles from an urn, and Spinning a Spinner) that work pretty much the same way. This section contains information common to these four simulators. If you'd like to skip this section for now and start using one of these simulators, by all means do so. The directions for the Settings editor in each simulator are explained later in this chapter; they refer you back to this section when needed.

Trial Set

Most Settings editors in the Probability Simulation application contain the **Trial Set** option. With this option, you tell the application how many trials you want it to run whenever you tell it to toss coins, roll dice, pick marbles, or spin a spinner. For example, if you are tossing two coins and **Trial Set** is 5, the first time you tell the application to toss the coins, it tosses the two coins five times, thus executing five trials. The next time you tell it to toss the coins, it tosses the coins five more times for a total of ten trials, and so on. Similarly, if you are rolling dice and **Trial Set** is 5, then each time you tell the application to roll dice, it will roll the dice five times.

If **Trial Set** is 1, then, after you tell the application to perform the first trial, you are given the option of tossing the coins one time, ten times, or 50 times. (You can see an example of this in the second picture of Figure 12-5, later in this chapter.) If **Trial Set** is greater than 1, then each time you tell the application to execute a trial, it executes the number of trials specified by the **Trial Set** setting.

Graph and table settings

Most of the Settings editors in the Probability Simulation application contain options that you can use to tell the application how you want it to display the graph and the table of the data it collects. These options are

- **Graph:** With this option, the application graphs your data as a histogram and gives you the choice between having the *y*-axis of the histogram display the *frequency* of an outcome or the *probability* of an outcome. For example, if you toss one coin eleven times and it lands heads up six times, then in a frequency histogram the bar depicting the "heads" outcome is six units tall; in a probability histogram, it is 6/11 units tall. When setting the options in the Settings editor, select **Freq** if you want the graph displayed as a frequency histogram; select **Prob** for a probability histogram.
- **StoTbl:** This option tells the application how many rows you want to see when your data is displayed in a table. You are given a choice between seeing no rows, all the rows, or only 50 rows. If you select 50 rows, only the bottom 50 rows of the table (your most recent data) can be viewed as you scroll through the data in a table. But let's face it, the TI-84 is a powerful puppy; it can handle displaying *all* the rows in the table. So there is no harm in setting the **StoTbl** option to **All**. It won't slow things down one bit.
- **ClearTbl:** This option is pretty obvious; it affords you the opportunity to tell the application if you want to clear the table or not. Your only choices are **Yes** and **No**. However, if you make changes to other options in the Settings editor, **ClrTbl** is automatically set to **Yes**.

- **Update:** This option enables you to tell the application to update the graph and table after 1, 20, or 50 trials, or at the end of the probability experiment. Because the TI-84 is a fast calculator, set this to 1.

Weighted outcomes

The Advanced Settings editor is where you tell the application if you want it to toss fair or weighted coins, roll fair or weighted dice, or spin a spinner divided into equal or unequal sections. The first picture in Figure 12-2 depicts the weights for a fair marble and the second picture shows the weights for a spinner divided into four equal sections.

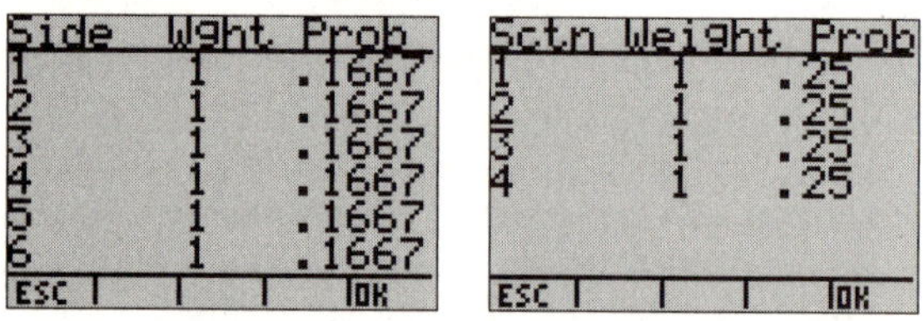

Evenly weighted marble Evenly divided spinner

Figure 12-2: Fair (unbiased) dice and spinner.

To bias a coin, die, or spinner, change the weight of the outcome. For example, to bias a die so that it is twice as likely to display 6 as it is to display any other number, set the weights of sides 1 through 5 to 1 and the weight of side 6 to 2.

To change the weights of the outcomes listed in the Advanced Settings editor, use the arrow keys to place the cursor on the existing weight in the second column of the Advanced Settings editor, use the keypad to enter the new weight, and then press [ENTER]. The number you enter for the weight must be a positive integer. As you enter the weights, the application updates the probability of the outcome in the third column.

Tossing Coins

The Probability Simulation application can simulate tossing from one to three coins at a time. It can even toss weighted coins. When tossing only one coin at a time, the application keeps track of the number of heads and tails that occur as the coin is repeatedly tossed. When tossing more than one coin at a time, the application keeps track of only the number of heads that occur.

The data from the repeatedly tossing coins can be displayed in a histogram or in a table. The application even gives you the option of saving your data so you can, if you choose, statistically analyze it at a later time. The "Saving your data" sidebar in this chapter tells you how to — you guessed it — save your data. This section tells you how to use the Toss Coins simulator.

Simulating the tossing of coins

To toss coins, follow these steps:

1. **Start the Probability Simulation application and press [1] to select the Toss Coins option in the Main menu.**

   You see the Toss Coins Home screen, as illustrated in the first picture in Figure 12-3.

   TIP: To use the Toss Coins simulator after using another simulator, repeatedly press [Y=] until you see the Main menu (second picture in Figure 12-1), and then press [1] to select the Toss Coins simulator.

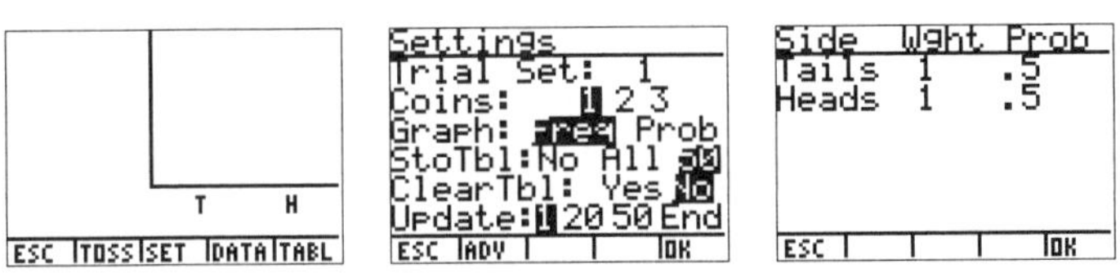

Home screen | Settings editor | Advanced Settings editor

Figure 12-3: The Toss Coins simulator's Home screen and editors.

2. **Press [ZOOM] to display the Settings editor. Change the settings in Settings editor as appropriate.**

   The second picture in Figure 12-3 shows the Settings editor. Should you need it, the "Understanding the Settings Editor" section, earlier in this chapter, explains the function of all options except the **Coins** option. The **Coins** option lets you tell the application to toss 1, 2, or 3 coins each time it tosses coins.

To enter a number in **Trial Set**, use the arrow keys to place the cursor on the existing number, use the keypad to enter a new positive integer, and then press [ENTER]. To select other options, use the arrow keys to place the cursor on the option, and then press [ENTER] to select the option.

3. **Press [WINDOW] to display the Advanced Settings editor. Enter the weights for your coins.**

   The screen you see is similar to the third picture in Figure 12-3. If you are tossing more than one coin, the application requires that all of the coins be weighted in the same way. For example, if you want one coin to be twice as likely to land heads up and you are tossing three coins, then all three coins must be twice as likely to land heads up. Should you need it, the "Understanding the Settings Editor" section, earlier in this chapter, explains how to weight your coins.

4. **Press [GRAPH] to say OK to the weights in the Advanced Settings editor. Press [GRAPH] again to say OK to the settings in the Settings editor.**

   You see the Toss Coins Home screen, as illustrated in the first picture in Figure 12-3.

5. **Press [WINDOW] to toss coins.**

   You see a screen similar to the first or second picture in Figure 12-4. If one coin was tossed only one time, then only one bar appears in the histogram.

6. **Continue tossing coins.**

   If in Step 2, you set **Trial Set** to a number greater than one, then after tossing the first set of coins, you see a screen similar to the first picture in Figure 12-4. Press [WINDOW] to toss more coins.

   If **Trial Set** is 1, then after tossing the first set of coins, you see a screen similar to the second picture in Figure 12-4. Press [WINDOW] to toss one more set of coins, [ZOOM] to toss ten more sets of coins, or [TRACE] to toss 50 more sets of coins.

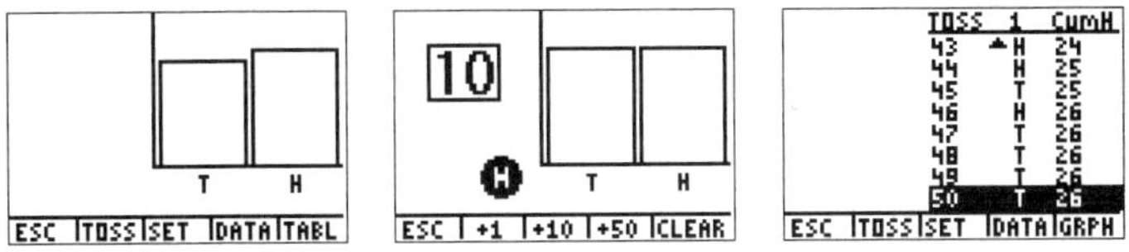

Figure 12-4: The Toss Coins simulator's graph and table.

Reading the graph and table

The cumulative data from tossing coins is stored in both a histogram and a table. The histogram, by default, is displayed on the screen without any values on the *y*-axis. The "Histograms and tables" sidebar in this chapter tells you how to display and read the histogram and the table. Here are some specifics about the Toss Coins histogram and table:

- **Toss Coins histogram:** If you are tossing only one coin, the histogram displays the cumulative outcomes of tails and heads from repeated tossing of the coin, as illustrated in the first and second pictures in Figure 12-4. If you are tossing more than one coin, the histogram displays only cumulative outcomes of heads. For example, if you are tossing two coins, the histogram displays the outcomes of no heads, one head, or two heads.
- **Toss Coins table:** The number of the toss appears in the first column of the table. If only one coin is tossed, the second column contains the outcome (H for heads or T for tails) for that coin and the last column contains the cumulative number of heads, as illustrated in the third picture of Figure 12-4. If more than one coin is tossed, the middle columns contain the outcome (H or T) for each coin and the last column contains the total number of heads for that particular toss of the coins.

Rolling Dice

The Probability Simulation application can simulate rolling from one to three dice at a time. It can even roll weighted dice. When rolling only one die at a time, the application keeps track of the number of times each number on the die lands up as the die is repeatedly rolled. When rolling more than one die

Histograms and tables

The Toss Coins, Roll Dice, Pick Marbles, and Spin Spinner simulators in the Probability Simulation application display cumulative data in both a histogram and a table. The histogram, by default, is displayed on the screen without any values on the *y*-axis.

To view the *y*-values of the histogram, press [▸]. The first bar in the histogram is shaded and the *y*-value is displayed. Continue to press [▸] to display the *y*-values of the other bars. Pressing [▸] after the last bar takes you back to the first bar in the histogram. Pressing [◂] can also be used to display the *y*-values of the bars in the histogram.

To display the table of the data, you must be in the simulator's Home screen. You know you are in the Home screen when **TABL** is the last command at the bottom of the screen, as illustrated in the first picture of Figure 12-4. If you are not in the Home screen, as illustrated in the second picture of Figure 12-4, press [Y=] to display the Home screen.

To display the table from the simulator's Home screen, press [GRAPH] to execute the **TABL** command. To display the histogram after displaying the table, press [GRAPH] again to execute the **GRPH** command.

at a time, the application keeps track of the sum of the numbers on the dice.

The data from the repeatedly rolling dice can be displayed in a histogram or in a table. The application even gives you the option of saving your data so you can, if you choose, statistically analyze it at a later time. The "Saving your data" sidebar in this chapter tells you how to — you guessed it — save your data. This section tells you how to use the Roll Dice simulator.

Simulating the rolling of dice

To roll dice, follow these steps:

1. **Start the Probability Simulation application and press [2] to select the Roll Dice option in the Main menu.**

   You see the Roll Dice Home screen, as illustrated in the first picture in Figure 12-5.

Tip: To use the Roll Dice simulator after using another simulator, repeatedly press [Y=] until you see the Main menu (second picture in Figure 12-1), and then press [2] to select the Roll Dice simulator.

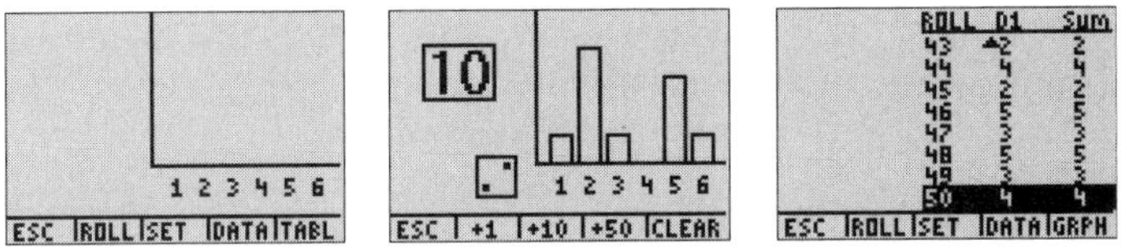

Home screen Graph Table

Figure 12-5: The Roll Dice simulator's Home screen, graph, and table.

2. **Press [ZOOM] to display the Settings editor. Change the settings in Settings editor as appropriate.**

   You see a screen similar to the Toss Coins simulator's Settings editor, as illustrated in the second picture in Figure 12-3.

   Should you need it, the "Understanding the Settings Editor" section, earlier in this chapter, explains the function of all options except the **Dice** and **Sides** options. The **Dice** option lets you tell the application to roll 1, 2, or t3 dice each time it rolls dice. The **Sides** option lets you roll normal six-sided dice or exotic dice having 8, 10, 12, or 20 sides. If you are rolling more than one die, all your dice must have the same number of sides.

   To enter a number in **Trial Set**, use the arrow keys to place the cursor on the existing number, use the keypad to enter a new positive integer, and then press [ENTER]. To select other options, use the arrow keys to place the cursor on the option and press [ENTER] to select the option.

   Don't forget the **Update** option. The down arrow in the lower-right corner of the screen indicates that you should repeatedly press [▼] to reach this option.

3. **Press [WINDOW] to display the Advanced Settings editor. Enter the weights for your dice.**

   The screen you see lists the sides of a die in the first column, its weight in the second column, and the probability of that side landing face up in the third column. If you are rolling more than one die, the application requires that all of the dice be weighted in the same way. Should you need it, the earlier section, "Understanding the Settings Editor," explains how to weight your dice.

4. **Press [GRAPH] to say OK to the weights in the Advanced Settings editor. Press [GRAPH] again to say OK to the settings in the Settings editor.**

   You see the Roll Dice Home screen, as illustrated in the first picture in Figure 12-5. If you are rolling more than one die, you see a jumble of numbers on the screen. This is explained in the next section, "Reading the graph and table."

5. **Press [WINDOW] to roll dice.**

   You see a screen similar to the second picture in Figure 12-5. If in Step 2 you set **Trial Set** to 1, then only one bar appears in the histogram. If you set **Trial Set** to a number greater than one, then the command line at the bottom of the screen is the same command line as in the first picture in this figure.

6. **Continue rolling dice.**

   If in Step 2 you set **Trial Set** to a number greater than one, press [WINDOW] to roll more dice.

   If **Trial Set** is 1, then after rolling the first set of dice, you see a screen similar to the second picture in Figure 12-5. Press [WINDOW] to roll one more set of dice, [ZOOM] to roll ten more sets of dice, or [TRACE] to roll 50 more sets of dice.

Reading the graph and table

The cumulative data from rolling dice is stored in both a histogram and a table. The histogram, by default, is displayed on the screen without any values on the *y*-axis. The "Histograms and tables" sidebar in this chapter tells you how to display and read the histogram and the table. Here are some specifics about the Roll Dice histogram and table:

- **Roll Dice histogram:** If you are rolling only one die, the histogram displays the cumulative outcomes of the side of the die that landed face-up. This is illustrated in the second picture in Figure 12-5 for a six-sided die. If you are rolling more than one die, the histogram displays cumulative outcomes of the sum of the face-up sides of the dice. For example, if you are rolling three dice, the possible outcomes (sums of face-up sides) are the integers 3 through 15. These numbers appear below the histogram with the two-digit numbers displayed in a column. The right arrow indicates that there are too many bars in the histogram to display on the screen. Repeatedly press [▶] to see these bars.

- **Roll Dice table:** The number of the roll appears in the first column of the table. The middle columns contain the outcome (number that landed face-up) for each die, and the last column contains the sum of the outcomes for the dice in that roll. If you are rolling only one die, the middle and last column will, of course, be the same, as illustrated in the third picture in Figure 12-5.

Saving your data

The data that appears in any Probability Simulation table can be saved to lists so that you can, after exiting the application, statistically analyze this data.

To save the data, go to the simulator's Home screen. (You know you are in the Home screen when **DATA** is the fourth command at the bottom of the screen. If you are not in the Home screen, press [Y=] to display the Home screen.) Then press [TRACE] to display the Save Data screen. The names of the files (lists) you see are approximately the same as the column headings in the simulator's table. You cannot rename these lists. Press [GRAPH] to tell the application that it is OK to save the data to the lists specified on the Save Data screen.

The list names for the data in each simulator are different. So, for example, if you save data after tossing coins and then save data after rolling dice, the data from each simulation will remain in the calculator. But the list names in a given simulator are the same. So, for example, if you save data after tossing coins and then save data again after tossing more coins, the data from the second simulation will — without warning — overwrite the data from the first simulation. If you don't want this to happen, consider renaming the lists saved in the first simulation before saving the data from the second simulation. To rename the lists, you must first exit the Probability Simulation application. Then create a user-named data list using a new list name and recall the old list to that user-named list. The user-named list is automatically saved under its list name.

Picking Marbles from an Urn

The Probability Simulation application can simulate picking one marble at a time from an urn that contains from two to five different types (colors) of marbles. You even get to specify how many marbles of each type are in the urn and you get to tell the application whether or not a marble picked from the urn is placed back in the urn before the next selection is made. The application keeps track of the type of marble picked from the urn.

The data from the repeatedly picking marbles can be displayed in a histogram or in a table. The application even gives you the option of saving your data so that you can, if you choose, statistically analyze it at a later time. The "Saving your data" sidebar in this chapter tells you how to — you guessed it — save your data. This section tells you how to use the Pick Marbles simulator.

Simulating picking marbles

To pick marbles from an urn, follow these steps:

1. **Start the Probability Simulation application and press [3] to select the Pick Marbles option in the Main menu.**

   You see the Pick Marbles Home screen, as illustrated in the first picture in Figure 12-6.

   TIP: To use the Pick Marbles simulator after using another simulator, repeatedly press [Y=] until you see the Main menu (second picture in Figure 12-1) and then press [3] to select the Pick Marbles simulator.

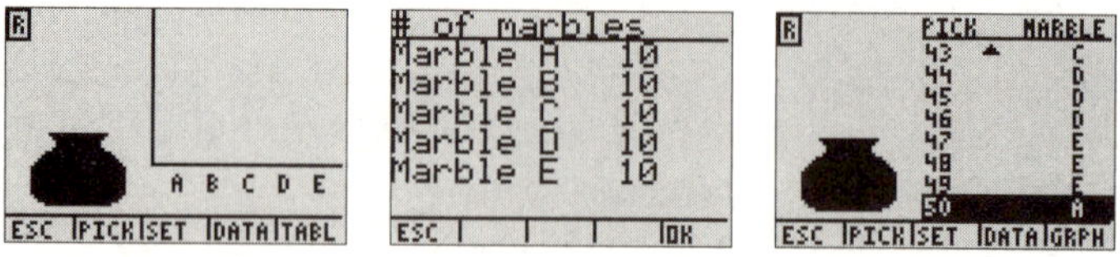

Home screen | Advance Settings screen | Table

Figure 12-6: The Pick Marbles simulator's Home screen, Advanced Settings screen, and table.

2. **Press ZOOM to display the Settings editor. Change the settings in Settings editor, as appropriate.**

   You see a screen similar to the Toss Coins simulator's Settings editor, as illustrated in the second picture in Figure 12-3.

   Should you need it, the earlier section, "Understanding the Settings Editor," explains the function of all options except the **Types** and **Replace** options. The **Types** option lets you tell the application how many different types (colors) of marbles are in the urn. The urn can contain from two to five different kinds of marbles. The **Replace** option lets you tell the simulator if a marble is put back in the urn (**Yes**) or not (**No**) before another marble is picked from the urn.

   To enter a number in **Trial Set**, use the arrow keys to place the cursor on the existing number, use the keypad to enter a new positive integer, and then press ENTER. To select other options, use the arrow keys to place the cursor on the option, and then press ENTER to select the option.

   Don't forget the **Update** option. The down arrow in the lower-right corner of the screen indicates that you should repeatedly press ▼ to reach this option.

3. **Press WINDOW to display the Advanced Settings editor. Enter the number of each type of marble in the urn.**

   The Advanced Settings editor is illustrated in the second picture in Figure 12-6. To change the number of marbles of a certain type, use the arrow keys to place the cursor on the number appearing next to that type, use the keypad to enter a new positive integer, and then press ENTER.

4. **Press GRAPH to say OK to the weights in the Advanced Settings editor. Press GRAPH again to say OK to the settings in the Settings editor.**

   You see the Pick Marbles Home screen. If you set the **Replace** option to **Yes**, the letter R appears in the upper-left corner of the screen, as illustrated in the first picture in Figure 12-6.

5. **Press [WINDOW] to pick marbles.**

   The histogram appears on the screen. If in Step 2 you set **Trial Set** to 1, then only one bar appears in the histogram.

6. **Continue picking marbles.**

   If in Step 2 you set **Trial Set** to a number greater than one, press [WINDOW] to pick more marbles.

   If **Trial Set** is 1, then after picking the first marble, the command line at the bottom of the screen is the same as the command line in the second picture in Figure 12-5. Press [WINDOW] to pick one more marble, [ZOOM] to pick ten more marbles, or [TRACE] to pick 50 more marbles.

   TIP: If in Step 2 you set **Replace** to **No**, then the urn could eventually run out of marbles. When this happens, the simulator displays the "no more marbles" message and will, of course, not allow you to pick any more marbles from the urn. To pick more marbles after seeing this message, start a new simulation of picking marbles by following Steps 2 through 6. In Step 2 if you don't see the **SET** command, press [Y=] to get back to the Pick Marbles simulator's Home screen.

Reading the graph and table

The cumulative data from picking marbles is stored in both a histogram and a table. The histogram, by default, is displayed on the screen without any values on the *y*-axis. The "Histograms and tables" sidebar in this chapter tells you how to display and read the histogram and the table. Here are some specifics about the Pick Marbles simulator's histogram and table:

- **Pick Marbles histogram:** Each bar in the histogram records the data for each type of marble in the urn.
- **Pick Marbles table:** This is illustrated in the third picture of Figure 12-6. The first column displays the number of the pick and the second column shows what type of marble was picked.

Spinning a Spinner

The Probability Simulation application can simulate spinning a spinner that is divided into two to eight sections of equal or unequal areas. It will not simulate the spinner landing on the boundary of two sections. An outcome of this simulator is the section on which the spinner landed.

The data from the repeatedly spinning spinner can be displayed in a histogram or in a table. The application even gives you the option of saving your data so you can, if you choose, statistically analyze it at a later time. The "Saving your data" sidebar in this chapter tells you how to — you guessed it — save your data. This section tells you how to use the Spin Spinners simulator.

Simulating spinning spinners

To spin a spinner, follow these steps:

1. **Start the Probability Simulation application and press [4] to select the Spin Spinners option in the Main menu.**

   You see the Spin Spinner simulator's Home screen, as illustrated in the first picture in Figure 12-7.

   To use the Spin Spinner simulator after using another simulator, repeatedly press [Y=] until you see the Main menu (second picture in Figure 12-1), and then press [4] to select the Spin Spinners simulator.

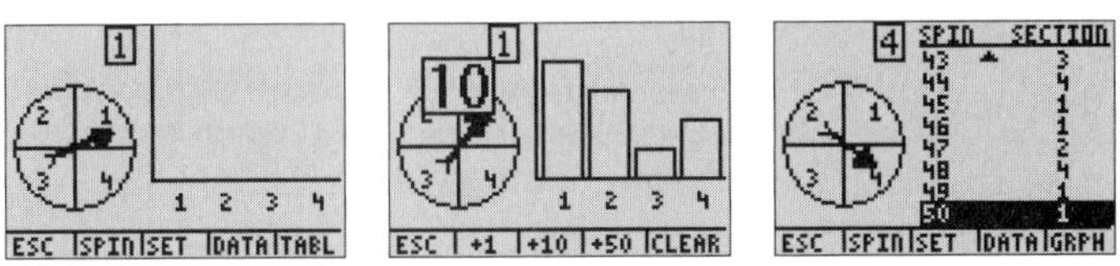

Home screen Graph Table

Figure 12-7: The Spin Spinner simulator's Home screen, graph, and table.

2. **Press [ZOOM] to display the Settings editor. Change the settings in Settings editor as appropriate.**

You see a screen similar to the Toss Coins simulator's Settings editor, as illustrated in the second picture in Figure 12-3.

Should you need it, the earlier section, "Understanding the Settings Editor," explains the function of all options except the **Sections** option. This option lets you tell the application how many sections you want the spinner to have. It can have from two to eight sections.

To enter a number in **Trial Set** or **Sections**, use the arrow keys to place the cursor on the existing number, use the keypad to enter a new positive integer, and then press [ENTER]. To select other options, use the arrow keys to place the cursor on the option and press [ENTER] to select the option.

3. **Press [WINDOW] to display the Advanced Settings editor. Enter the weight of each section of the spinner.**

   The number of the section is displayed in the first column of the Advanced Settings editor, its weight appears in the second column, and the probability of the spinner landing on that section appears in the third column. The weight corresponds to the area of the spinner occupied by that section. For example, if you want a spinner to have three sections with one section occupying half the area of the spinner and the other two sections each occupying a fourth of the spinner, set the weight of one section to two and the weight of the other two sections to one.

   To change a number in the weight column, use the arrow keys to place the cursor on the existing number, use the keypad to enter a new positive integer, and then press [ENTER].

4. **Press [GRAPH] to say OK to the weights in the Advanced Settings editor. Press [GRAPH] again to say OK to the settings in the Settings editor.**

   You see the Spin Spinner Home screen.

5. **Press [WINDOW] to spin the spinner.**

   You see a screen similar to the second picture in Figure 12-7. The number on which the spinner last landed appears at the top of the screen.

6. **Continue spinning the spinner.**

   If in Step 2 you set **Trial Set** to a number greater than one, press [WINDOW] to spin the spinner the number of times specified in **Trial Set**.

   If **Trial Set** is 1, then after the first spin of the spinner the command line at the bottom of the screen is the same as the command line in the second picture in Figure 12-7. Press [WINDOW] to spin the spinner one time, [ZOOM] to spin the spinner ten more times, or [TRACE] to spin it 50 more times.

Reading the graph and table

The cumulative data from spinning the spinner is stored in both a histogram and a table. The histogram, by default, is displayed on the screen without any values on the *y*-axis. The "Histograms and tables" sidebar in this chapter tells you how to display and read the histogram and the table. Here are some specifics about the Spin Spinner simulator's histogram and table:

- **Spin Spinner histogram:** Each bar in the histogram records the data for each section of the spinner.
- **Spin Spinner table:** This is illustrated in the third picture of Figure 12-7. The first column displays the number of the spin and the second column shows where the spinner landed.

Drawing Cards from a Deck

The Probability Simulation application can simulate drawing a card from one to three decks of cards. It even allows you to specify whether or not the card is placed back in the deck before another card is drawn. And those decks of cards can consist of 52 or 32 cards. A deck of 32 cards is a 52-card deck without any twos, threes, fours, fives, or sixes.

The application keeps track of the card picked each time the simulation is run and the data is displayed in a table. The application even gives you the option of saving your data so you can, if you choose, statistically analyze it at a later time.

The "Saving your data" sidebar in this chapter tells you how to — you guessed it — save your data. This section tells you how to use the Draw Cards simulator.

Simulating drawing cards

To draw cards from one or more decks of cards, follow these steps:

1. **Start the Probability Simulation application and press [5] to select the Draw Cards option in the Main menu.**

   You see the **Draw Cards** Home screen, as illustrated in the first picture in Figure 12-8.

   TIP: To use the **Draw Cards** simulator after using another simulator, repeatedly press [Y=] until you see the Main menu (second picture in Figure 12-1), and then press [5] to select the **Draw Cards** simulator.

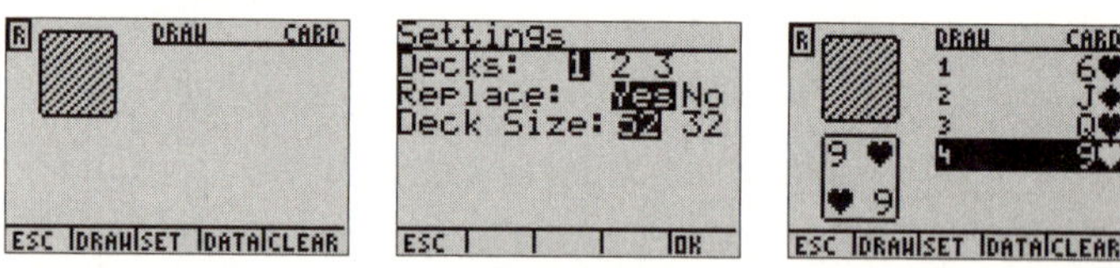

Home screen Settings editor Table

Figure 12-8: The **Draw Cards** simulator's Home screen, Settings editor, and table.

2. **Press [ZOOM] to display the Settings editor. Change the settings in Settings editor, as appropriate.**

   You see the screen in the second picture in Figure 12-8. **Decks** tells the application how many decks of cards you want to use, **Replace** tells the application if you want the card placed back in the deck or decks of cards before another card is drawn, and **Deck Size** tells the application how many cards are in the deck or decks of cards you are using. To select these options, use the arrow keys to place the cursor on the option and press [ENTER] to select the option.

3. **Press [GRAPH] to say OK to the settings in the Settings editor.**

   You see the Draw Cards Home screen. If you set the **Replace** option to **Yes**, the letter R appears in the upper-left corner of the screen, as illustrated in the first picture in Figure 12-8.

4. **Press [WINDOW] to draw cards.**

   Each time you press [WINDOW], the card drawn is displayed in the table on the screen, as illustrated in the third picture in Figure 12-8.

TIP

If in Step 2 you set **Replace** to **No**, then the deck could eventually run out of cards. When this happens the simulator displays the "no more cards" message and will, of course, not allow you to draw any more cards. When this happens you have three options: Go do something else, press [GRAPH] to clear the table and start over with the same type of simulation, or press [ZOOM] to set up a different type of simulation.

Saving the Draw Cards simulator's data

The "Saving your data" sidebar in this chapter tells you how to save the data stored in any table appearing in the Probability Simulation application. When you save the data from the Draw Cards simulator, it is saved in three lists: Draw, Value, and Suit. The Draw list records the number of the draw. The Value list records the numerical value of the card. The numerical value of an Ace is 1, a Jack is 11, a Queen is 12, and a King is 13. The Suit list records a Heart as 1, a Club as 2, a Spade as 3, and a Diamond as 4.

Chapter 13

Using GeoMaster

In This Chapter

- Starting and quitting GeoMaster
- Creating, saving, recalling, and deleting files
- Using menus
- Clearing the screen

If you've ever used a dynamic geometry computer software package such as the Geometer's Sketchpad or Cabri Geometry II Plus, then you appreciate its ability to not only accurately construct geometric figures, but to also perform translations, rotations, dilations, and reflections. These software packages can even take measurements, create tessellations, and much, much more. If you haven't used such software packages, wouldn't you like to?

Because the TI-84 comes equipped with geometry applications, including GeoMaster, you can now do your geometric investigations on the calculator. Granted, the calculator screen is rather small for some investigations, but other than that, these two calculator applications do almost everything that the computer software packages do.

In this and the following two chapters, I explain how to use the GeoMaster application.

When you start GeoMaster, you see the version number on the screen, as illustrated in the first picture in Figure 13-1. If that number is smaller than 1.10, the version in existence at the time this book was written, then check out Texas Instruments' Web site to see if there is a more recent version for you to download.

Starting GeoMaster

To start GeoMaster, press [APPS][ALPHA][TAN], if necessary use [▼] to move the cursor to the GeoMaster application, and press [ENTER] to select the application. You are confronted with one of the three screens in Figure 13-1.

GeoMaster screen | GeoMaster is running | Another application is running

Figure 13-1: What you see when you select the GeoMaster application.

If no other applications are running, you see the first screen in Figure 13-1. Just press any key to enter the GeoMaster application. If GeoMaster is already running, after you press [APPS], you see the second screen in Figure 13-1. Press [2] to re-enter the application. The third screen in Figure 13-1 tells you that another application is currently running. Press [2] to exit that application, and then press any key to enter GeoMaster. After you enter GeoMaster, you are confronted with one of the two screens in Figure 13-2.

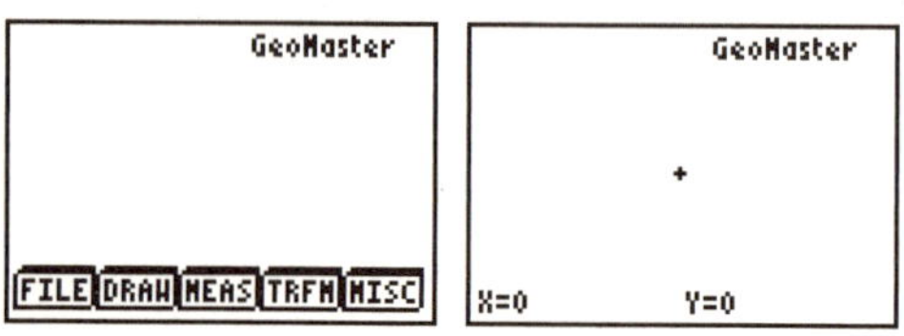

With menus | Without menus

Figure 13-2: What you see when you enter or re-enter GeoMaster.

When you enter the GeoMaster application, you usually see the first screen in Figure 13-2. But if you re-enter GeoMaster while it is still running, you see the screen as you left it but without the menus at the bottom of the screen, as in the second screen in Figure 13-2. To reinstate the Menu headings at the bottom of the screen, press [GRAPH] or [ALPHA].

If you press a menu key such as [MODE], [STAT], [MATH], or [VARS] while GeoMaster is running, you are sent to that menu while GeoMaster continues to run in the background. The same thing happens when you press [Y=], [WINDOW], or [ZOOM] when the Menu bar is not at the bottom of the GeoMaster screen. To re-enter the GeoMaster application, just press [GRAPH] or [APPS][2].

Quitting GeoMaster

You can use either a clean way or a dirty way of quitting the GeoMaster application. The clean way truly terminates GeoMaster, *without saving your work;* the dirty way leaves GeoMaster running in the background. When you re-start GeoMaster after a dirty exit, your work will be the way you left it — provided you don't run any other applications in the interim.

Before performing a clean exit from GeoMaster, you may want to save your work since this type of exit eradicates anything that you have not saved. How to save your work is explained in the next section. A clean exit is achieved by using the **Quit** command in GeoMaster's **FILE** menu ([Y=][6][1]).

A dirty exit, which leaves GeoMaster running in the background, is achieved by pressing [2nd][MODE] or any other key combination that sends you elsewhere — such as [MODE] which sends you to the Mode menu. After performing a dirty exit, you re-enter GeoMaster by pressing [GRAPH] or [APPS][2]. To reinstate the menu headings at the bottom of the screen, press [GRAPH] or [ALPHA].

When you start GeoMaster, the application removes the highlights from all equal signs in the Y= editor so that no function will be graphed when you press the [GRAPH] key to, for example, display GeoMaster's menu bar. It also resets the viewing window. When you exit GeoMaster, the application leaves all the equal signs in the Y= editor not highlighted and does not return the viewing window to its previous settings. For more information on the Y= editor and the viewing window, refer to Chapter 8.

Creating and Saving Files

You can create a file in two ways. If the GeoMaster screen is empty, just start creating your geometric figures. If the GeoMaster screen is not empty, select **New File** from GeoMaster's File menu by pressing [Y=][1] if the Menu bar is visible or [GRAPH][Y=][1] if it is not. You will be asked if you want to save your previous work. If you select **Yes**, the calculator is automatically put in Alpha mode so that you can enter a filename. Press [ENTER] after entering the filename. (Entering filenames is explained in Step 3 in the following list.) After selecting **Yes** and entering a filename or after selecting **No**, you are confronted with a blank GeoMaster screen, on which you can start creating geometric figures. Creating Geometric figures is explained in Chapter 14.

To save your work, follow these steps:

1. **Select Save File from GeoMaster's File menu.**

   To do this, press [Y=][3] if the Menu bar is visible; press [GRAPH][Y=][3] if it is not. After selecting **Save File**, you see a screen similar to the first picture in Figure 13-3.

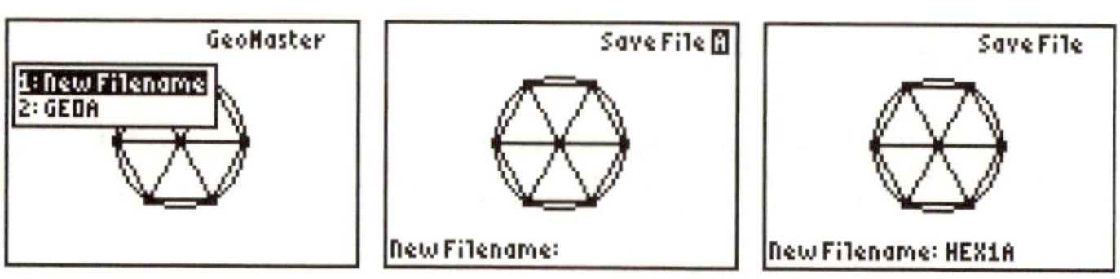

Press [Y=] [3] — Select **New Filename** — Enter a name

Figure 13-3: The steps for saving a GeoMaster file.

2. **Select an existing filename or create a new filename.**

   If you select an existing filename, such as GEOA in the first picture in Figure 13-3, the contents of that file will be replaced with your new work. If you select **New Filename**, you see a screen similar to the second picture in Figure 13-3.

   GeoMaster does not allow you to save a blank screen.

3. **If you selected New Filename, give your file a name and then press [ENTER].**

The name of your file can consist of one to eight characters, which must be letters, numbers, or the Greek letter θ. The first character in the name must be a letter or θ.

The [a] in the upper-right corner of the screen indicates that the calculator is in Alpha mode. So when you press a key you will be entering the green letter above the key. To enter a number, press [ALPHA] to take the calculator out of Alpha mode and then enter the number. To enter a letter after entering a number, you must press [ALPHA] to put the calculator back in alpha mode. This is illustrated in the third picture in Figure 13-3.

Press [ENTER] to save the file under its new name. As the file is being saved, you see a moving vertical line in the upper-right corner of the screen. This indicates that the calculator is busy saving your file.

WARNING! If you save a GeoMaster file under the name of an existing GeoMaster file, the original file will be replaced by the new file. The calculator will give you no warning message that this is going to happen.

Recalling and Deleting Files

To recall a GeoMaster file previously saved in memory, select **Open File** from GeoMaster's **FILE** menu by pressing [Y=][2] if the Menu bar is visible or [GRAPH][Y=][2] if it is not. Then press the number of the file you want to recall.

To delete a GeoMaster file from the memory of the calculator, follow these steps:

1. **Press [2nd][+][2] to enter the Memory manager.**

   It may take a few seconds for the Memory manager to appear.

2. **Repeatedly use [▼] until the cursor is to the left of AppVars, and then press [ENTER].**

3. **Repeatedly use [▼][▲] until the cursor is to the left of the file you want to delete and press [DEL].**

4. **Press [2] if you really want to delete the file; press [1] if you chicken out about deleting the file from memory.**

Using Menus

When you enter GeoMaster, the Menu bar appears at the bottom of the screen. If you are re-entering the application after it has been running in the background, press [GRAPH] or [ALPHA] to make the Menu bar appear.

The five menus on GeoMaster's Menu bar are

- **FILE:** This menu contains the commands that allow you to create, save, and recall files. How to do this is explained in the sections, "Creating and Saving Files" and "Recalling and Deleting Files," earlier in this chapter. The menu also contains the command that allows you to clear the screen and quit the GeoMaster application.
- **DRAW:** The tools housed in this menu help you draw geometric figures. How to use these tools is explained in Chapter 14.
- **MEAS:** This menu contains the commands that allow you to take measurements (length, distance, angle, area, and so on). Using these commands is explained in Chapter 15.
- **TRFM:** If you want to transform objects by translating, reflecting, rotating, and/or dilating them, the tools needed to perform these tasks are found in this menu.
- **MISC:** This menu contains some really great commands that answer the questions: Are these points collinear? Are these lines parallel or perpendicular? It also houses the commands for hiding objects in your construction. How to use these commands is explained in Chapter 14.

TIP: If you are cruising through GeoMaster's menus and decide not to make a selection, press [CLEAR] to get out of the menus and back to the GeoMaster screen.

To select a menu, press the key under the menu. That is, press [Y=] to enter the **FILE** menu, [WINDOW] to enter the **DRAW** menu, [ZOOM] to enter the **MEAS** menu, [TRACE] to enter the **TRFM** menu, and [GRAPH] to enter the **MISC** menu. A down arrow after the number of the last item in a menu screen or an up arrow after the number of the first menu item on the screen indicates that there are more items in the menu than can be displayed on the screen. This is illustrated in the two pictures in Figure 13-4. Use [▼][▲] to view these items.

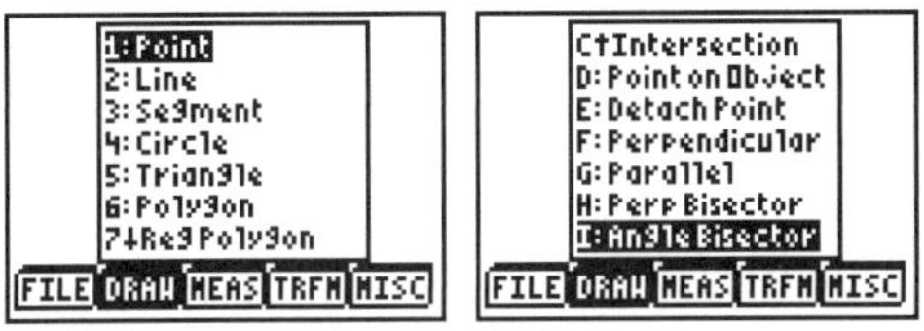

More items are below More items are above

Figure 13-4: A menu containing more items than can fit on one screen.

To activate a menu item, use [▼][▲] to move the cursor to the desired item and press [ENTER]. Using the commands housed in these menus is explained in the next two chapters.

TIP: If you press [▲] when the cursor is at the top of the menu, as in the first picture in Figure 13-4, the cursor jumps to the last menu item, as illustrated in the second picture in Figure 13-4. If you press [▼] when the cursor is at the bottom of a menu, the cursor jumps to the top of the menu.

Clearing the Screen

The **Clear All** command in GeoMaster's **FILE** menu erases all objects on the screen. It is equivalent to starting a new file without saving your work. But be warned, GeoMaster does not have an undo feature. So if you clear the screen, what was on the screen is gone for good. To select **Clear All** from the **FILE** menu, press [Y=][5] if the Menu bar is visible or [GRAPH][Y=][5] if it is not.

TIP

[GRAPH] and [CLEAR] are your friends

Can't find GeoMaster's Menu bar?

Press [GRAPH] and it instantly appears.

Did you accidentally exit GeoMaster?

Press [GRAPH] and it appears as you left it.

Want to get rid of the Menu bar?

Press [CLEAR] and it vanishes.

Want to stop using a Menu item?

Press [CLEAR] and the item is deactivated.

Want to get out of the Menu bar without selecting an item?

Press [CLEAR] and you're out.

Change your mind about creating a new file?

Press [CLEAR] and no new file will be created.

In a nutshell, [GRAPH] gets you back into GeoMaster while it is running in the background and displays the menu bar while it is running in the foreground. [CLEAR] stops whatever it was you were doing or it makes the Menu bar disappear.

Chapter 14

Constructing Geometric Figures

In This Chapter

- Constructing geometric figures
- Finding midpoints, points of intersection, and angle bisectors
- Determining if two lines are parallel or perpendicular
- Determining if three points are collinear
- Hiding, deleting, and moving objects
- Adjusting GeoMaster's viewing window

The GeoMaster **DRAW** menu contains a plethora of tools that enable you to do mundane things such as construct lines, segments, triangles, polygons, and circles. The **DRAW** menu also contains tools that enable you to do really neat things such as bisecting angles, constructing perpendicular and parallel lines, locating midpoints, and finding the points of intersection of geometric objects. And the **MISC** menu contains commands that will get the calculator to tell you if two lines are parallel or perpendicular, or if three points are collinear. This chapter tells you how to use these commands.

Constructing Points

This section explains how to construct a freestanding point, that is, a point that is not intended to be part of another object such as a point on a line or a point on the circumference of a circle. Constructing a point that is part of an existing object is explained later in this chapter.

TIP

Precisely locating and labeling points

You can enter the coordinates of the location of the point after pressing [ENTER] to approximately establish the location of the point. But be warned, the numbers you enter for these coordinates can contain no more than one decimal place. If you enter a number with two or more decimal places for the *x*-coordinate, the calculator will not allow you to enter the *y*-coordinate. To enter the coordinates of the point after pressing [ENTER] to approximately locate the point, use the number keys to enter the single-decimal-place *x*-coordinate of the point and press [ENTER]. Then use the number keys to enter the *y*-coordinate of the point and press [ENTER]. The point is constructed at the specified coordinates.

Points can be labeled immediately after they are constructed. For example, if you are constructing a segment, after pressing [ENTER] to establish the first point defining that segment, you can label it before continuing with the construction of the segment. To do this, after pressing [ENTER] to establish the point, press [ALPHA] to label it with one letter, or press [2nd][ALPHA] to label it with more than one letter. Enter the label and continue with your construction.

You can tell the calculator where you want to construct a point in two ways: Use the arrow keys to identify the location of the point, or tell the calculator the coordinates of the point. Both ways are described in the following steps. To construct a freestanding point, follow these steps:

1. **Press [GRAPH][WINDOW][1] to select the Point tool from thc DRAW menu.**

   "Point" appears in the upper-right corner of the screen.

2. **Use the arrow keys to move the cursor to the location where you want to construct the point and press [ENTER].**

   As you move the cursor, the coordinates of its location appear at the bottom of the screen. The point is constructed at the location of the cursor.

   TIP

   After pressing [ENTER] in this step, you can precisely locate the point by giving the calculator the coordinates of that point. You can also label the point. The

"Precisely locating and labeling points" sidebar in this chapter tells you how to do this.

3. **Press [ENTER] or move the cursor away from the object.**

   This is a crucial step. If it is not followed, then the last point you constructed may vanish when you ask GeoMaster to perform another operation.

After constructing one point, the **Point** tool remains active until you deactivate it. So if you want to construct more points, just repeat Steps 2 and 3. When you're finished using the **Point** tool, deactivate it by pressing [CLEAR] or by selecting another menu item.

Constructing Lines, Segments, Rays, and Vectors

Because two points determine a line, segment, ray, or vector, the directions for creating these objects are pretty much the same. To construct such objects, follow these steps:

1. **Select the appropriate tool from the DRAW menu.**

   To do this, press [WINDOW] if the Menu bar is visible at the bottom of the screen or [GRAPH][WINDOW] if it is not. Use [▼] to move the cursor to the tool you wish to use and press [ENTER] to select that tool. The tool you selected appears in the upper-right corner of the screen. The first picture in Figure 14-1 illustrates that the **Ray** tool was selected from the **DRAW** menu.

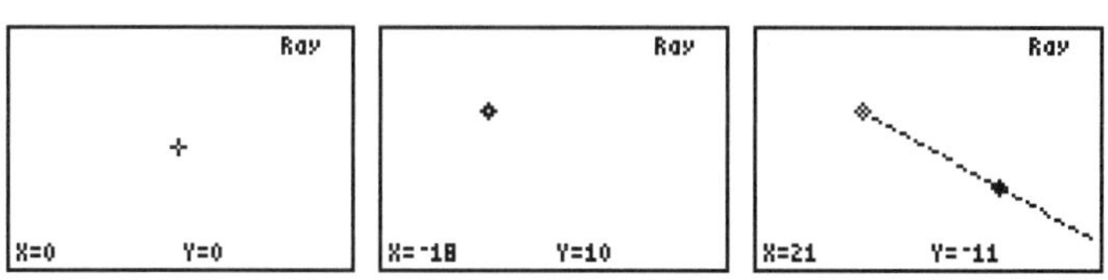

Select **Ray** tool Anchor first point Anchor second point

Figure 14-1: Constructing a ray. (When done, press [ENTER].)

2. **Use the arrow keys to move the cursor to the location of the first point that determines your line, segment, ray, or vector. Press [ENTER] to establish this point.**

As you move the cursor, its coordinates are displayed at the bottom of the screen and the anchored point appears in the shape of a diamond, as illustrated in the second picture in Figure 14-1.

After pressing [ENTER] in this step, you can precisely locate the point by giving the calculator the coordinates of that point. You can also label the point. The "Precisely locating and labeling points" sidebar in this chapter tells you how to do this.

3. **Use the arrow keys to move the cursor to the location of the second point that determines the object and press [ENTER] to anchor that point.**

   The third picture in Figure 14-1 illustrates this step when constructing a ray. Notice that the ray appears as a dashed line. This indicates that the ray, although it appears on the screen, has not yet been housed in the memory of the calculator.

4. **Press [ENTER] or move the cursor away from the object.**

   This is a crucial step. If it is not followed, then the last object you created may vanish when you ask GeoMaster to perform another operation. After doing this step, the figure that appeared as a dashed line after performing Step 3 appears as a solid line. This indicates that your construction is now housed in the memory of the calculator.

After constructing a line segment, ray, or vector, the GeoMaster tool you were using remains active until you deactivate it. So if you want to construct more objects of the same kind, just repeat Steps 2 through 4. When you are finished using the tool, you deactivate it by pressing [CLEAR] or by selecting another menu item.

Constructing Circles

A circle is determined by its center and its radius. To construct a circle, follow these steps:

1. **Select the Circle tool from the DRAW menu.**

   To do this, press [GRAPH][WINDOW][4]. "Circle" in the upper-right corner of the first picture in Figure 14-2 illustrates that the **Circle** tool is active.

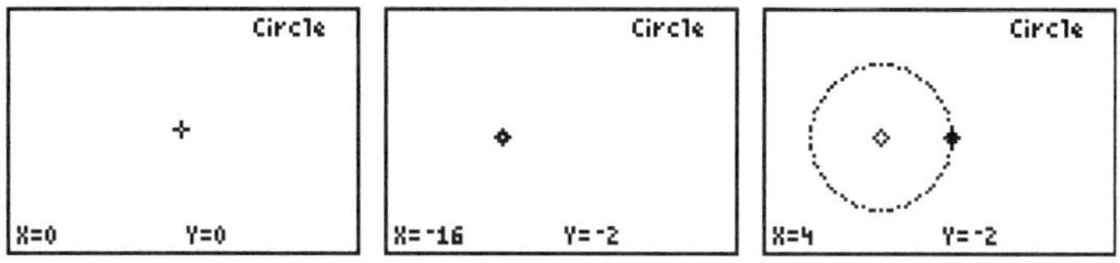

Select **Circle** tool   Anchor the center   Anchor the size

Figure 14-2: Constructing a circle.

2. **Use the arrow keys to move the cursor to the location on the screen where you want to place the center of the circle. Press [ENTER] to anchor this point.**

   As you move the cursor, its coordinates are displayed at the bottom of the screen and the center of the circle appears in the shape of a diamond, as illustrated in the second picture in Figure 14-2.

   After pressing [ENTER] in this step, you can precisely locate the center by giving the calculator the coordinates of that point. You can also label the center. The "Precisely locating and labeling points" sidebar in this chapter tells you how to do this.

3. **Use the arrow keys to expand the circle to the desired size and press [ENTER] to anchor the size of the circle.**

   The third picture in Figure 14-2 illustrates this step. The circle appears as dashed lines, indicating that the construction is not yet complete.

   After pressing [ENTER] in this step, you can precisely locate the point on the circumference by giving the calculator the coordinates of that point. You can also label this. The "Precisely locating and labeling points" sidebar in this chapter tells you how to do this.

4. **Press [ENTER] or move the cursor away from the object.**

   This is a crucial step. If it is not followed, then the last circle you created may vanish when you ask GeoMaster to perform another operation. After doing this step, the circle which appeared as a dashed circle after performing Step 3 appears as a solid circle. This indicates that your construction is now housed in the memory of the calculator.

Did you make a mistake?

If you make a mistake immediately after completing a construction (or part of a construction), press [CLEAR] to erase the construction, and then start over. This will erase the whole construction made by the active tool appearing in the upper-right corner of the screen. But that tool remains active, so you can start over with the same construction. If you need to erase a previously constructed object, the "Hiding and Deleting Objects" section, later in this chapter, tells you how to do this.

After constructing your first circle, the **Circle** tool remains active until you deactivate it. So if you want to construct more circles, just follow Steps 2 through 4. When you are finished using this tool, you deactivate it by pressing [CLEAR] or by selecting another menu item.

After a circle has been constructed, you can move its location on the screen by moving the circumference of the circle, and you can change the size of its radius by moving the point on the circumference of the circle. Moving points and objects is explained later in this chapter.

Constructing Triangles and Arcs

A triangle and an arc are determined by three points, so their construction is pretty much the same. To construct a triangle or an arc, follow these steps:

1. **Select the appropriate tool from the DRAW menu.**

   To do this, press [WINDOW] if the Menu bar is visible at the bottom of the screen or [GRAPH][WINDOW] if it is not. Use [▼] to move the cursor to the tool you want to use and press [ENTER] to select that tool. The tool you selected appears in the upper-right corner of the screen. The first picture in Figure 14-3 illustrates that the **Arc** tool was selected from the **DRAW** menu.

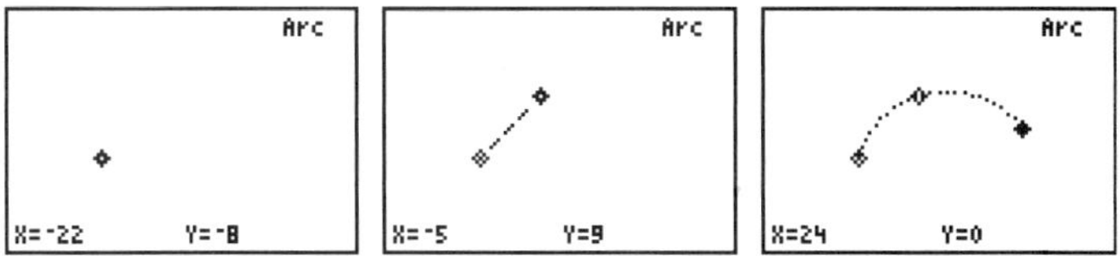

Anchor first point   Anchor second point   Anchor third point

Figure 14-3: Constructing an arc. (When done, press [ENTER].)

2. **Use the arrow keys to move the cursor to the location on the screen where you want to construct the first point of the triangle or arc. Press [ENTER] to construct this point.**

   As you move the cursor, its coordinates are displayed at the bottom of the screen and the anchored point appears in the shape of a diamond, as illustrated in the first picture in Figure 14-3.

   After pressing [ENTER] in this step, you can precisely locate the point by giving the calculator the coordinates of that point. You can also label the point. The "Precisely locating and labeling points" sidebar in this chapter tells you how to do this.

3. **Use the arrow keys to move the cursor to the location on the screen where you wish to construct the second point defining the triangle or arc. Press [ENTER] to anchor this point.**

   The second picture in Figure 14-3 illustrates this step when constructing an arc.

4. **Use the arrow keys to move the cursor to the location on the screen where you wish to construct the third point defining the triangle or arc. Press [ENTER] to anchor this point.**

   The third picture in Figure 14-3 illustrates this step when constructing an arc. The arc appears as dashed lines indicating that the construction is not yet complete.

5. **Press [ENTER] or move the cursor away from the object.**

This is a crucial step. If it is not followed, then the last object you created may vanish when you ask GeoMaster to perform another operation. After doing this step, the figure which appeared as dashed lines or arcs after performing Step 4 appears as solid lines or arcs. This indicates that your construction is now housed in the memory of the calculator.

After constructing a triangle or arc, the GeoMaster tool you were using remains active until you deactivate it. So if you want to construct more objects of the same kind, just follow Steps 2 through 5. When you are finished using that tool, you deactivate it by pressing [CLEAR] or by selecting another menu item.

After a triangle or an arc has been constructed, you can move its location on the screen by moving one of the segments defining the triangle or arc, or you can change its shape by moving the points defining the object. Moving points and objects is explained later in this chapter.

Constructing Polygons

If you want to create a regular polygon, the next section tells you how to do it. To create an (irregular) polygon, follow these steps:

1. **Press [GRAPH][WINDOW][6] to select the Polygon tool from the DRAW menu.**

   "Polygon" in the upper-right corner of the screen indicates that the **Polygon** tool is active.

2. **Use the arrow keys to move the cursor to the location on the screen where you want to construct the first point defining your polygon. Press [ENTER] to anchor this point.**

   As you move the cursor, its coordinates are displayed at the bottom of the screen, and the anchored point appears in the shape of a diamond, as illustrated in the first picture in Figure 14-4.

   After pressing [ENTER] in this step, you can precisely locate the point by giving the calculator the coordinates of that point. You can also label the point. The "Precisely locating and labeling points" sidebar in this chapter tells you how to do this.

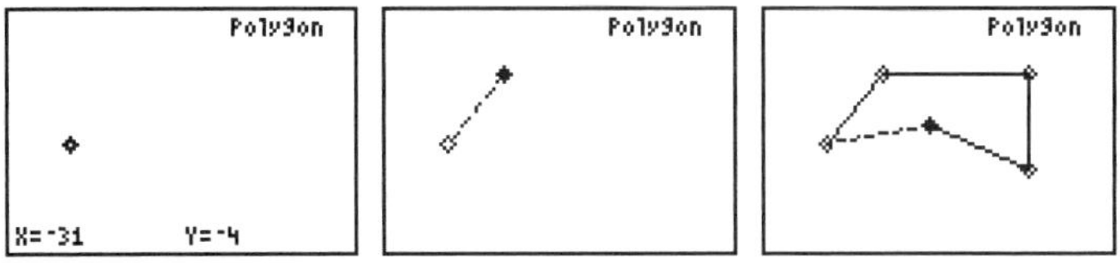

Anchor first point — Anchor all except last point — Anchor last point

Figure 14-4: Constructing a polygon.

3. **Use the arrow keys to move the cursor to the location on the screen where you wish to construct the second point defining the polygon. Press [ENTER] to anchor this point.**

   The second picture in Figure 14-4 illustrates this step. Notice that the dashed line between these points indicates that the polygon is not yet constructed.

4. **Use the arrow keys to construct the remaining points that define the polygon. Press [ENTER] to anchor each point.**

   The third picture in Figure 14-4 illustrates this step. Notice that the line joining the first and last points you constructed is a dashed line. This indicates that the construction of the polygon is not yet complete.

5. **Press [ENTER] and move the cursor away from the object.**

   All sides of the polygon appear as solid lines indicating the construction of the polygon is complete.

After constructing a polygon, the **Polygon** tool remains active until you deactivate it. So if you want to construct more polygons, just follow Steps 2 through 5. When you are finished using this tool, you deactivate it by pressing [CLEAR] or by selecting another menu item.

Constructing Regular Polygons

A regular polygon is determined by its center and the number of its sides. To construct a regular polygon, follow these steps:

1. **Press [GRAPH][WINDOW][7] to select the Reg Polygon tool from the DRAW menu.**

   "Reg Polygon" in the upper-right corner of the screen indicates that the **Reg Polygon** tool is active.

2. **Use the arrow keys to move the cursor to the location on the screen where you want to place the center of the regular polygon. Press [ENTER] to anchor this point.**

   As you move the cursor, its coordinates are displayed at the bottom of the screen.

   TIP: After pressing [ENTER] in this step, you can precisely locate the center by giving the calculator the coordinates of that point. You can also label the point. The "Precisely locating and labeling points" sidebar in this chapter tells you how to do this.

3. **Press [▸] to display the number of sides of the polygon. The default number of sides is 6. Press [−] to decrease the number of sides of the polygon, or press [+] to increase the number of sides. *Do not* press [ENTER] when your polygon has the desired number of sides or you will be unable to rotate or adjust the size of the polygon.**

   This is illustrated in the first picture in Figure 14-5 for the construction of a pentagon. GeoMaster can construct polygons having a minimum of 3 sides and a maximum of 12 sides.

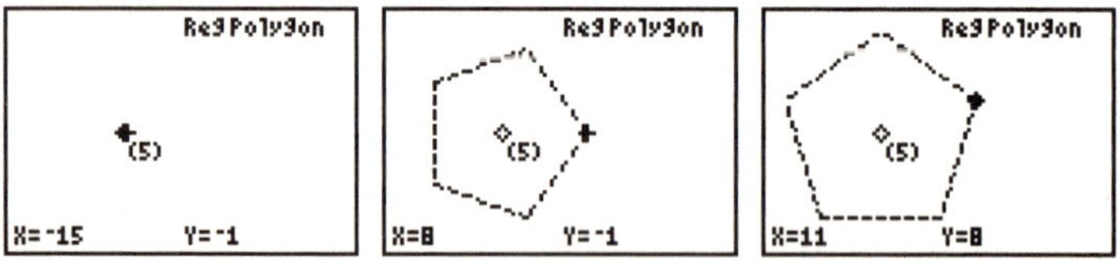

Anchor center and size — Adjust size — Rotate and adjust size

Figure 14-5: Constructing a regular polygon.

4. **Repeatedly press [◂] and/or [▸] to adjust the size of your polygon and repeatedly press [▴] and/or [▾] to rotate the polygon. Press [ENTER] *only when* the polygon is to your liking.**

Pressing [►] increases its size and [◄] decreases its size. This is illustrated in the second picture of Figure 14-5. Pressing [▲] rotates the polygon in the counterclockwise direction and [▼] rotates it in the clockwise direction. Continue using these arrow keys, in any order, until the polygon is exactly the way you want it. Press [ENTER] when you are done. This is illustrated in the third picture in Figure 14-5.

5. **Press [ENTER] or move the cursor away from the polygon.**

   This is a crucial step. If it is not followed, then the last polygon you created may vanish when you ask GeoMaster to perform another operation. After doing this step, the polygon which appeared as dashed lines after performing Step 4 appears as solid lines. This indicates that your construction is now housed in the memory of the calculator.

After constructing your first regular polygon, the **Reg Polygon** tool remains active until you deactivate it. So if you want to construct more regular polygons, just repeat Steps 2 though 5. When you are finished using this tool, you deactivate it by pressing [CLEAR] or by selecting another menu item.

If you want to label the vertices defining your regular polygon, the "Labeling objects" sidebar in this chapter tells you how to do this.

Constructing Perpendicular and Parallel Lines

The next section tells you how to construct a perpendicular bisector. This section tells you how to construct perpendicular or parallel lines. To construct such lines, you have to start with an already constructed line, segment, ray, vector, or side of a polygon or triangle. Earlier sections in this chapter explain how to construct these objects. To construct a line that is perpendicular or parallel to an already constructed line, segment, ray, vector, or side of a triangle or polygon, follow these steps:

1. **Select the appropriate tool from the DRAW menu.**

   To do this, press [WINDOW] if the Menu bar is visible at the bottom of the screen or [GRAPH][WINDOW] if it is not. Use [▼] to move the cursor to the tool you wish to use and press [ENTER] to select that tool. The tool you select appears in the upper-right corner of the screen. The first picture in Figure 14-6 illustrates that the **Perpendicular** tool was selected from the **DRAW** menu.

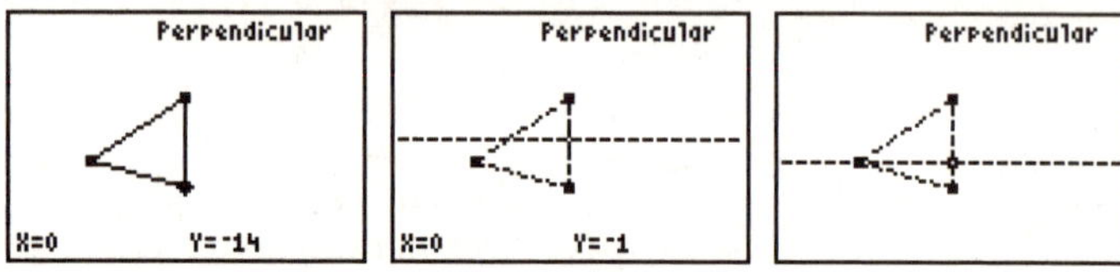

Select **Perpendicular** tool   Select given line   Move perpendicular

Figure 14-6: Constructing a perpendicular line. (When done, press [ENTER].)

2. **Use the arrow keys to move the cursor to the already constructed line, segment, ray, vector, or side of a triangle or polygon, and then press [ENTER].**

   A perpendicular or parallel line appears, as illustrated in the second picture in Figure 14-6. In the next step, this line is moved to the location of your choice.

3. **Use the arrow keys to move the cursor to the location on the screen where you want to construct the perpendicular or parallel line. Press [ENTER] to anchor this location.**

   The third picture in Figure 14-6 illustrates this step when constructing a perpendicular line. Notice that the coordinates of the cursor location appear at the bottom of the screen and that your figure is drawn in dashed lines, indicating that the construction is not yet complete.

4. **Press [ENTER] or move the cursor away from the object.**

   This is a crucial step. If it is not followed, then the last object you created may vanish when you ask GeoMaster to perform another operation. After doing this step, the figure which appeared as a dashed line after performing Step 3 appears as a solid line. This indicates that your construction is now housed in the memory of the calculator.

After constructing a parallel or perpendicular line, the GeoMaster tool you were using remains active until you deactivate it. So if you want to construct more objects of the same kind, just repeat Steps 2 though 4. When you are finished using that tool, you deactivate it by pressing [CLEAR] or by selecting another menu item.

Constructing Perpendicular Bisectors

In GeoMaster, a perpendicular bisector can be constructed to an already existing line, segment, ray, vector, or side of a polygon or triangle. Constructing these objects is explained earlier in this chapter. To construct a perpendicular bisector, follow these steps:

1. **Select the Perp Bisector tool from the DRAW menu.**

   To do this, press [WINDOW] if the Menu bar is visible at the bottom of the screen or [GRAPH][WINDOW] if it is not. Press [▲] to move the cursor to the bottom of the menu, press [▲] again to move the cursor to the **Perp Bisector** tool, and press [ENTER]. The tool you selected appears in the upper-right corner of the screen.

2. **Use the arrow keys to move the cursor to the already constructed line and press [ENTER].**

   The perpendicular bisector appears as a solid line. This indicates that you are finished with the construction.

After constructing a perpendicular bisector, the **Perp Bisector** tool remains active until you deactivate it. So if you want to construct more perpendicular bisectors, just repeat Step 2. When you are finished using the tool, you deactivate it by pressing [CLEAR] or by selecting another menu item.

Constructing Angle Bisectors

For GeoMaster to recognize the existence of an angle, that angle must contain three defining points — the vertex and a point on each side of the angle. Also, GeoMaster recognizes a point only if it appears on the screen as a small square, such as the three points defining the triangle in Figure 14-6. If the angle you want to bisect does not contain the small squares representing the three points defining the angle, then GeoMaster cannot bisect it. But all is not lost. The next section tells you how to establish the point of intersection of two lines, and the section after that tells you how to place a point on a line. After you have constructed the three points defining your angle, GeoMaster can bisect it.

To construct the bisector of an already constructed angle containing three defining points, follow these steps:

1. **Select the Angle Bisector tool from the DRAW menu.**

   To do this, press [WINDOW] if the Menu bar is visible at the bottom of the screen or [GRAPH][WINDOW] if it is not. Press [▲] to move the cursor to **Angle Bisector**, the last menu item, and press [ENTER]. The tool you selected appears in the upper-right corner of the screen.

2. **Use the arrow keys to move the cursor to an existing point on one side of the angle and press [ENTER] to anchor that point.**

   GeoMaster uses a small square to indicate that a point exists. If you don't see a small square on the side of the angle, the first paragraph in this section tells you what to do about the situation.

3. **Use the arrow keys to move the cursor to the vertex of the angle and press [ENTER] to anchor the vertex.**

   The vertex must always be the second point that defines the angle.

4. **Use the arrow keys to move the cursor to an existing point on the other side of the angle and press [ENTER] to anchor that point.**

   The angle bisector appears as a solid line, indicating that your construction is complete.

After constructing an angle bisector, the **Angle Bisector** tool remains active until you deactivate it. So if you want to construct more angle bisectors, just repeat Steps 2 through 4. When you are finished using the tool, you deactivate it by pressing [CLEAR] or by selecting another menu item.

Constructing Points of Intersection

GeoMaster can construct the points of intersection of any pair of the following types of objects: lines, segments, rays, vectors, sides of polygons, circles, and arcs. If there is more than one point of intersection, GeoMaster will find all of these points.

To construct points of intersection, follow these steps:

1. **Select the Intersection tool from the DRAW menu.**

   To do this, press [WINDOW] if the Menu bar is visible at the bottom of the screen or [GRAPH][WINDOW] if it is not. Repeatedly press [▼] to highlight the **Intersection** tool and press [ENTER]. The tool you selected appears in the upper-right corner of the screen.

2. **Use the arrow keys to move the cursor to one of the objects in the intersecting pair of objects and press [ENTER].**

   After pressing [ENTER], the object just selected appears as a dashed figure.

3. **Use the arrow keys to move the cursor to the other object in the intersecting pair and press [ENTER].**

   The point or points of intersection appear on the screen and the **Intersection** tool remains active. You can construct more points of intersections by following Steps 2 and 3. When you are finished using this tool, you deactivate it by pressing [CLEAR] or by selecting another menu item.

TIP: When you press [ENTER] to select an object and GeoMaster isn't sure which object you want, it gives you a list of objects from which you choose one by pressing the number of the object. This is illustrated in Figure 14-7.

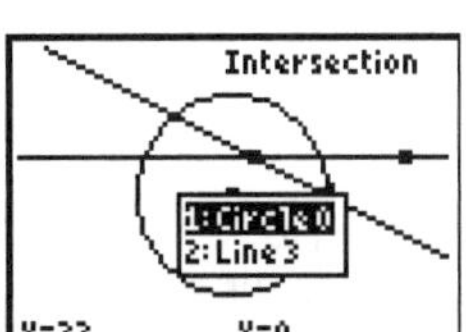

Figure 14-7: What you see when GeoMaster doesn't know what you want.

TIP: If you want to label the point of intersection you just constructed, the "Labeling objects" sidebar in this chapter tells you how to do this.

Constructing a Point on an Object

The first section in this chapter explains how to construct a freestanding point; this section explains how to construct a point so that it is part of an existing object. To construct a point on an object, first select the **Point on Object** tool from the **DRAW** menu (press [GRAPH][WINDOW], repeatedly press [▼] to highlight the **Point on Object** tool, and press [ENTER]). Then use the arrow keys to move the cursor to the location on the object where you wish to place a point and press [ENTER]. The point is constructed at the location of the cursor and the **Point on Object** tool remains active so that you can construct more points on the same object or another object. When you are finished constructing points on objects, press [CLEAR] to deactivate the **Point on Object** tool or select another menu item.

The **Detach Point** tool in the **DRAW** menu is used to detach a point that was constructed using the **Point on Object** tool. The detached point remains on the screen, but GeoMaster no longer considers it to be associated with the object to which it was previously attached. So if you move or transform the object, though the object changes, the point will not change its location. When the **Detach Point** tool is active, you detach a point by placing the cursor on the point and pressing [ENTER].

TIP: If you want to label the point you just constructed, the "Labeling objects" sidebar in this chapter tells you how to do this.

Constructing a Midpoint

GeoMaster can construct the midpoint of an already constructed segment or the midpoint between two existing points. The segment can be the side of a triangle or polygon. GeoMaster recognizes only those points that appear on the screen as small squares.

To construct a midpoint, first select the **Midpoint** tool from the **DRAW** menu (press GRAPH WINDOW, repeatedly press ▼ to highlight the **Midpoint** tool, and press ENTER). If you are constructing the midpoint of a segment, use the arrow keys to place the cursor on the segment and press ENTER to construct the midpoint. If you are constructing the midpoint between two points, move the cursor to the first point and press ENTER. Then move the cursor to the second point and press ENTER to have GeoMaster construct the midpoint.

After constructing the first midpoint, the **Midpoint** tool remains active so that you can construct more midpoints. When you are finished constructing midpoints, press CLEAR to deactivate the **Midpoint** tool or select another menu item.

TIP

If you want to label the midpoint you just constructed, the "Labeling objects" sidebar in this chapter tells you how to do this.

Moving Figures

When you move an object, all other objects associated with that object will also undergo a change. Figure 14-8 provides a surprising example. The circle and line in the first picture of this figure are associated objects because the center of the circle is also one of the points defining the line. So if you move *only* the line, as illustrated in the second picture of Figure 14-8, the center of the circle moves with the line because it is on the line. But the point on the circumference of the circle remains fixed because it is not on the line. Because the center and point on the circumference determine the radius of the circle, the size of the circle changes when only the line is moved.

And if you move *only* the circle, as illustrated in the third picture in Figure 14-8, the slope of the line changes because one of the points defining the line (the point that is also the center of the circle) moves with the circle but the other point stays fixed. Now if you move *both* the circle and the line, then the size of the circle and the slope of the line remain the same. This is illustrated in the third picture in Figure 14-9.

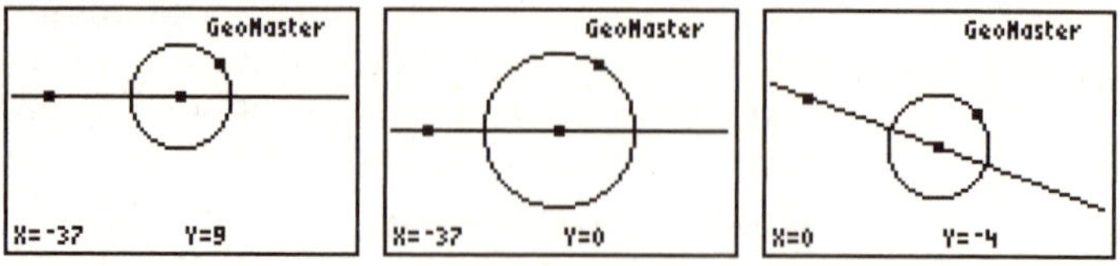

Associated objects | Circle moved | Line moved

Figure 14-8: Moving associated objects.

TIP

To reshape a polygon or triangle, move one or more of its vertices. To change the size of a circle, move a point on the circumference of that circle or move the center of the circle.

To move one or more objects, follow these steps:

1. **Press [CLEAR] to deactivate all GeoMaster tools.**

   When no tools are active, you see "GeoMaster" in the upper-right corner of the screen, as illustrated in the first picture in Figure 14-8.

2. **Use the arrow keys to move the cursor to the first object you want to move and press [ENTER].**

   The object you selected is displayed as dashed lines. This is illustrated in the first picture in Figure 14-9, where the circle is one of the objects that are to be moved.

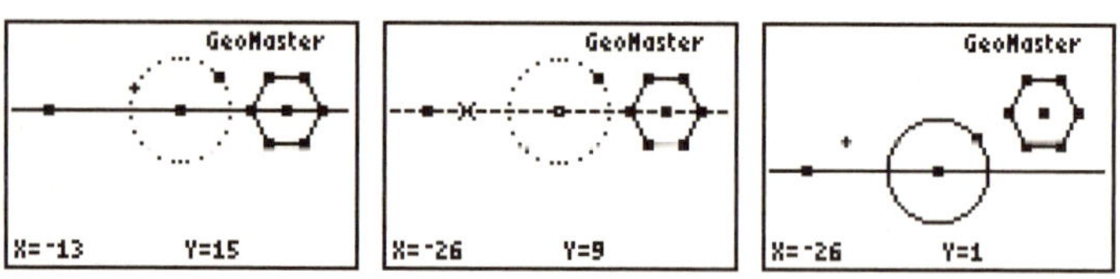

Select first object | Select other objects | Move objects

Figure 14-9: Moving one or more objects.

3. **If you are moving more than one object, move the cursor to the second object and press [2nd][ENTER]. Select other objects in the same manner — move the cursor to the object and press [2nd][ENTER].**

The additional objects you select appear as dashed lines. This is illustrated in the second picture in Figure 14-9, where the line was selected as the other object to be moved.

4. **When you are finished selecting objects to be moved, press [ENTER] while the cursor is on one of the selected objects.**

   The cursor appears in the shape of a times sign, as illustrated in the second picture in Figure 14-9.

5. **Use the arrow keys to move the selected objects and press [ENTER] to anchor the new location.**

   The cursor changes to the shape of a small plus sign.

6. **Move the cursor to a blank spot on the screen and press [ENTER].**

   The objects appear as solid lines, as illustrated in the third picture in Figure 14-9.

TIP

Labeling objects

I personally prefer to avoid labeling objects. Labels take time to create, crowd the screen, are rarely placed where you want them, and after you have gone to the trouble of moving them to a desirable location, they move out of place when you move or transform the object they label. But let's face it — there are times when you just have to label those points and objects so that you can refer to them by name. Figure 14-11 is a good example of a figure requiring labels.

To label an already constructed object, first press [GRAPH][GRAPH][2] to select the **Label** tool from the **MISC** menu. Then move the cursor to the point or object you want to label and press [ENTER] to select that object. a appears in the upper-right corner, indicating that the calculator is in Alpha mode so that you can enter letters. *Without moving the cursor,* enter the label for the object and press [ENTER] when you are finished. After creating one label, you can move the cursor to another object, press [ENTER], give that object a label, and press [ENTER] to anchor the label. When you are finished creating labels, deactivate the **Label** tool by pressing [CLEAR] or by selecting another menu item.

The label can contain up to eight characters consisting of letters and numbers. (Writing words in Alpha mode is explained in Chapter 1.) After you have created a label, you can move that label to a more desirable location. Moving objects, such as labels, is explained elsewhere in this chapter.

Determining if Two Lines are Parallel or Perpendicular

It is often the case that when you construct lines, two lines may look parallel or perpendicular, when in fact they aren't. The first picture in Figure 14-10 provides an example. Two lines in this picture appear to be parallel, and thus perpendicular to the third line, but they are not. How do I know this? I asked GeoMaster!

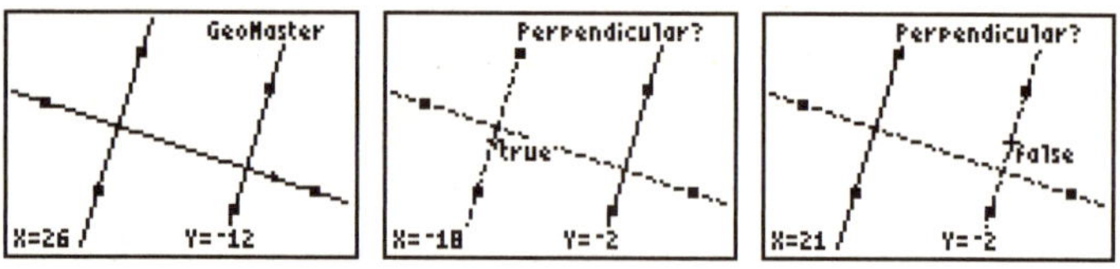

Given lines Two perpendicular lines Non-perpendicular lines

Figure 14-10: Determining if lines are perpendicular.

To get GeoMaster to determine if two lines are perpendicular or parallel, follow these steps:

1. **Select the Perpendicular? or Parallel? tool from the MISC menu.**

   To do this, press [GRAPH][GRAPH], use the arrow keys to move the cursor to the appropriate tool, and press [ENTER]. The tool you selected appears in the upper-right corner of the screen, as is illustrated in the second picture in Figure 14-10 where the **Perpendicular?** tool is active.

2. **Use the arrow keys to move the cursor to the first line and press [ENTER]. Move the cursor to the second line and press [ENTER]. The conclusion as to whether or not the lines are parallel or perpendicular appears on the screen.**

   The two selected lines appear as dashed lines, as illustrated in the second picture in Figure 14-10. The word "true" appearing on the screen in this picture indicates that the lines are indeed perpendicular.

3. **If you want to use the same tool to test another pair of lines, press [CLEAR] and then repeat Step 2.**

This is illustrated in the third picture in Figure 14-10. The word “false” appearing on the screen indicates that the lines are not perpendicular.

4. **When you are finished using the Perpendicular? or Parallel? tool, deactivate it by pressing CLEAR *two times* or by selecting another menu item.**

Determining if Three Points Are Collinear

Three points are collinear if they are on the same line. It is sometimes difficult to determine if three points are collinear by just looking at the screen. For example, points D, B, and E in the first picture in Figure 14-11 are collinear, but the points A, B, and C are not. How do I know this? I asked GeoMaster!

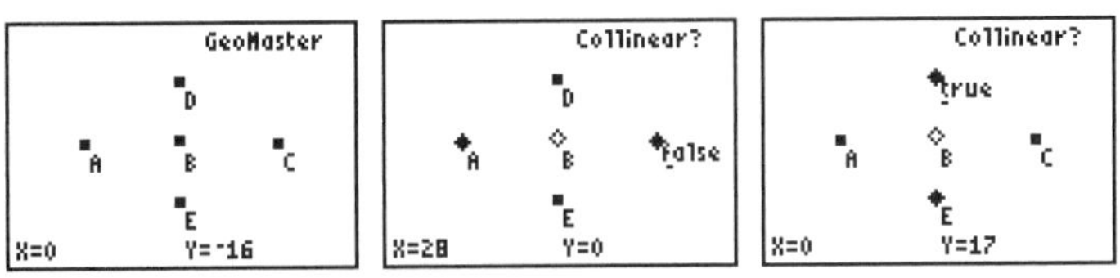

Given points Non-collinear points Collinear points

Figure 14-11: Determining if three points are collinear.

To determine if three points are collinear, follow these steps:

1. **Select the Collinear? tool from the MISC menu.**

   To do this, press GRAPH GRAPH, use the arrow keys to move the cursor to the **Collinear?** tool, and press ENTER. The tool you selected appears in the upper-right corner of the screen, as is illustrated in the second picture in Figure 14-11.

2. **Use the arrow keys to move the cursor to the first point and press ENTER. Move the cursor to the second point and press ENTER. Then move the cursor to the third point and press ENTER. The conclusion as to whether or not the points are collinear appears on the screen.**

   The selected points appear as diamonds, as illustrated in the second picture in Figure 14-11 where the points

A, B, and C are selected. The word "false" appearing on the screen in this picture indicates that the points are not collinear.

3. **If you want to determine if three other points are collinear, press CLEAR and then repeat Step 2.**

   This is illustrated in the third picture in Figure 14-11. The word "true" appearing on the screen indicates that the points D, B, and E are collinear.

4. **When you are finished using the Collinear? tool, deactivate it by pressing CLEAR *two times* or by selecting another menu item.**

Hiding and Deleting Objects

There is a big difference between hiding and deleting an object. If you hide an object, you can, if you so choose, bring it back in view at a later time. But if you delete an object, it is gone for good.

Hiding an object

The first picture in Figure 14-12 provides an example of when you may want to hide an object. Because GeoMaster has no tool that instantly creates a tangent to a circle, the tangents in this picture were created by first constructing radii, and then constructing lines perpendicular to these radii. (Constructing such objects is explained earlier in this chapter.) If all you really want to see is the circle and tangents, you can hide the radii, as illustrated in the third picture in Figure 14-12. If you were to delete the radii, the tangents would also be deleted because they were constructed perpendicular to the radii, and are thus dependent on the existence of the radii.

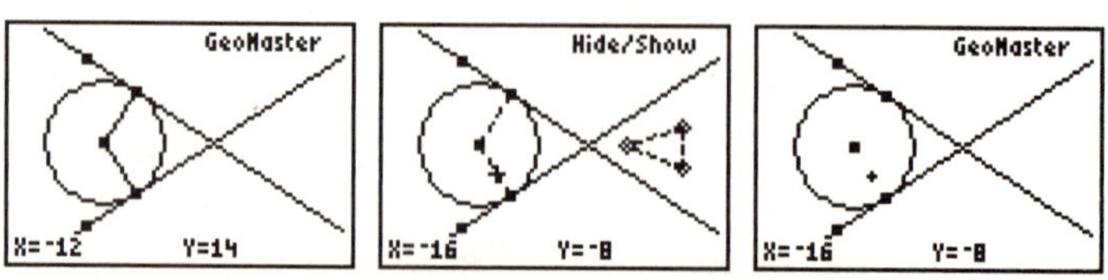

Given figure | Select objects to hide | Press CLEAR

Figure 14-12: Hiding objects.

To hide one or more objects, follow these steps:

1. **Press [GRAPH][GRAPH][▲][▲][▲][ENTER] to select the Hide/Show tool from the MISC menu.**

   The name **Hide/Show** appears in the upper-right corner of the screen and any previously hidden objects appear on the screen as dashed lines. This is illustrated in the second picture in Figure 14-12, where the triangle is a previously hidden object. For now, ignore any previously hidden objects. They will be hidden again in Step 3.

2. **Use the arrow keys to move the cursor to the first object you want to hide and press [ENTER]. Move the cursor to the second object you want to hide and press [ENTER]. Continue selecting objects to hide by placing the cursor on the object and pressing [ENTER].**

   The selected objects appear as dashed lines, as illustrated in the second picture in Figure 14-12, where the radii are selected.

3. **Press [CLEAR] to hide the objects selected in Step 2 and to again hide any previously hidden objects.**

   This is illustrated in the third picture in Figure 14-12, where the selected radii and the previously hidden triangle are all hidden.

Redisplaying hidden objects

To redisplay hidden objects, press [GRAPH][GRAPH][▲][▲][▲][ENTER] to select the Hide/Show tool from the MISC menu. All previously hidden objects appear on the screen. For each object that you no longer want to hide, place the cursor on that object and press [ENTER]. Then press [CLEAR]. The objects that you did not select to bring out of hiding are again hidden, and the objects you did select to show remain on the screen.

Deleting objects

WARNING!

GeoMaster has no undo feature. So if you delete an object, it's gone for good. Also, if you delete an object, then all other objects whose construction is dependent on the deleted object will also be deleted. For example, the tangents in the first picture in Figure 14-12 were constructed perpendicular to the radii of the circle. If you were to delete the radii, the tangents would also vanish because their existence is dependent on the existence of the radii.

To delete an object, press [CLEAR] to ensure that no tools are active. Use the arrow keys to move the cursor to the object and press [ENTER] to select that object. The object appears as a dashed line. Press [DEL] to delete the object.

Adjusting the Viewing Window

GeoMaster uses a default window in which $-47 \le x \le 47$ and $-31 \le y \le 31$. You can change this window by pressing [CLEAR][WINDOW] and redefining the settings in the Window editor. You can also change the window by pressing [CLEAR][ZOOM] and selecting one of the predefined Zoom features. The Window editor and the Zoom features are explained in Chapters 8 and 9. To return GeoMaster to its default window, press [CLEAR][ZOOM][8] to select **ZoomCenter** from the **Zoom** menu.

Chapter 15

Finding Measurements

In This Chapter

- Evaluating length and area
- Measuring an angle
- Constructing the sum of two vectors
- Determining the coordinates of a point or the slope of a line
- Finding the equation of a line or a circle

GeoMaster's **MEAS** menu houses the tools you use to find the length of a segment, degree measure of an angle, or area of a circle, triangle, or polygon. It also contains a tool that allows you to add and subtract like measurements. It even has a tool that determines the equation of a line or a circle. This chapter explains how to use these and other tools.

Finding Length and Distance

The **Distance/Length** tool in the **MEAS** menu is used to find the distance between two points, the length of a segment or vector, the circumference of a circle, or the perimeter of a polygon. To use this tool, follow these steps:

1. **Press [GRAPH][ZOOM][ENTER] to select the Distance/Length tool from the MEAS menu.**

   The name of the tool appears in the upper-right corner of the screen, as illustrated in the first picture of Figure 15-1.

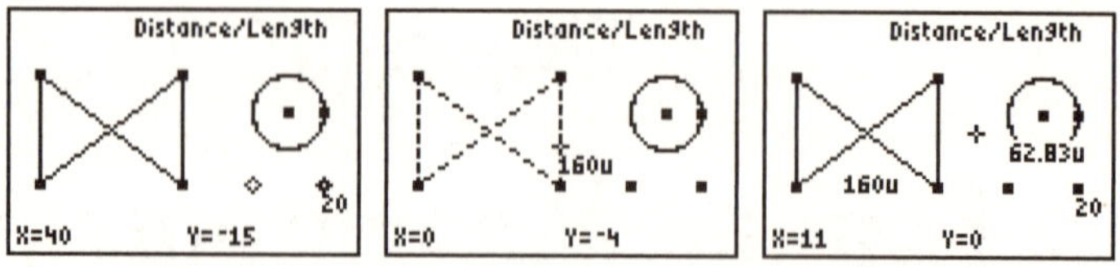

Distance Perimeter Circumference

Figure 15-1: Using the **Distance/Length** tool.

2. **Select the object you want to measure.**

   If you are measuring the distance between two points, use the arrow keys to move the cursor to the first point and press [ENTER] to select it. Then move the cursor to the second and press [ENTER]. The distance between the points appears on the screen, as illustrated in the first picture in Figure 15-1.

   If you are measuring the length of a segment or vector, the circumference of a circle, or the perimeter of a polygon, use the arrow keys to move the cursor to that object and press [ENTER] to select the object. The object is displayed using dashed lines and the requested measure appears on the screen. This is illustrated in the second picture in Figure 15-1, where GeoMaster found the perimeter of the polygon.

   TIP: If you don't like GeoMaster's placement of the length or distance, before performing the next step, you can use the arrow keys to relocate it. This is illustrated in the third picture in Figure 15-1 where the measurement of the perimeter of the polygon is relocated. The measurements of the other two objects in this picture remain where GeoMaster placed them.

3. **Press [ENTER] to anchor the placement of the measurement.**

4. **To measure another object, repeat Steps 2 and 3.**

   This is illustrated in the third picture of Figure 15-1, where all three objects on the screen have been measured.

5. **When you are finished taking measurements, press [ENTER] again or move the cursor away from the last object. To deactivate the Distance/Length tool, press [ENTER] or select another menu item.**

After pressing [ENTER], all objects are displayed using solid lines, as illustrated in the third picture in Figure 15-1.

Finding Area

GeoMaster can find the area of any circle, triangle, or *simple* polygon constructed using the **Circle**, **Triangle**, **Polygon**, or **RegPolygon** tool in the **DRAW** menu. (Using these tools is explained in Chapter 14.) A polygon is *simple* if its sides intersect only at the vertices of the polygon. For example, the polygon ABCD in Figure 15-3 is not simple, whereas all regular polygons, such as the hexagon in Figure 15-4, are simple.

In addition, GeoMaster is also not capable of directly finding the area of a polygon formed by the intersection of two other polygons, such as the quadrilateral in Figure 15-4 formed by the intersection of the hexagon and triangle. But all is not lost. The following subsections tell you how to find the area of a polygon that is not simple and the area of a polygon formed by the intersection of two other polygons.

To find the area of a circle, triangle, or simple polygon constructed using the tools in the **DRAW** menu, follow these steps:

1. **Press [GRAPH][ZOOM][2] to select the Area tool.**

   The name of the tool appears in the upper-right corner of the screen, as illustrated in Figure 15-2.

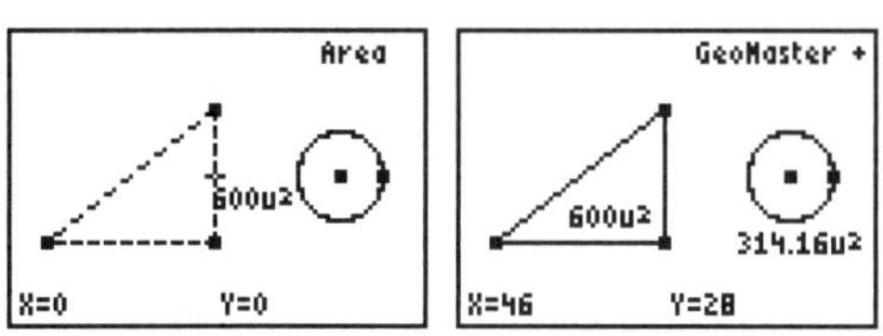

Area of triangle Area of circle

Figure 15-2: Finding the area of a triangle and a circle.

2. **Use the Arrow keys to move the cursor to the object whose area you want to find, and then press [ENTER] to select that object.**

The object is displayed using dashed lines and the requested area appears on the screen. This is illustrated in the first picture in Figure 15-2 for the area of the triangle.

TIP

If you don't like GeoMaster's placement of the area, before performing the next step, you can use the arrow keys to relocate it. This is illustrated in the second picture of Figure 15-2 for the location of the area of the triangle.

3. **Press [ENTER] to anchor the placement of the area measurement.**
4. **To find the area of another object, repeat Steps 2 and 3.**
5. **When you are finished finding areas, press [ENTER] again or move the cursor away from the last object. To deactivate the Area tool, press [ENTER] or select another menu item.**

   After pressing [ENTER], all objects are displayed using solid lines, as illustrated in the second picture in Figure 15-2.

Finding the area of a polygon having intersecting sides

To find the area of a polygon having intersecting sides, such as polygon ABCD in Figure 15-3, break the polygon down into simple polygons (polygons whose sides intersect only at vertices) and then add up the areas of the simple polygons.

Sounds simple, but it's not. To get GeoMaster to find the area of the smaller polygons, those polygons have to be constructed using the **Polygon** tool in the **DRAW** menu. But to construct a polygon using the **Polygon** tool, GeoMaster needs to know all the vertices of the polygon. So you must first find the point of intersection of the intersecting sides of the original polygon.

Again: Sounds simple, but it's not. Unfortunately, GeoMaster's **Intersection** tool can't find the intersection of sides of polygons. So here's what you can do to find the area of a polygon having intersecting sides:

1. **Press [GRAPH][WINDOW][3] to select the Segment tool from the DRAW menu. Construct segments over the sides of the intersecting sides of the polygon.**

   Using the **Segment** tool is explained in Chapter 14. In the polygon in Figure 15-3, I constructed the segments AD and BC.

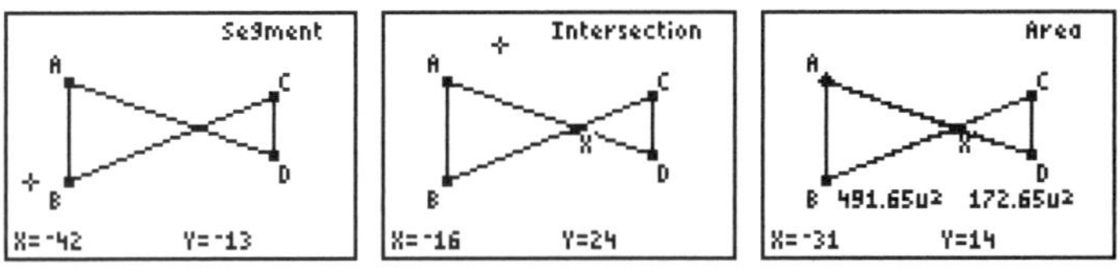

Construct segments | Find intersections | Construct polygons; find areas

Figure 15-3: Finding the area of a polygon having intersecting sides.

2. **Press [GRAPH][WINDOW][ALPHA][PRGM] to select the Intersection tool from the DRAW menu and use this tool to find the point of intersection of the segments constructed in Step 1.**

   Using the **Intersection** tool is explained in Chapter 14. If GeoMaster asks you which segment you want to select, choose the one with the highest number because this is the number of the most recently constructed segment. The second picture in Figure 15-3 shows that I labeled the point of intersection. You don't have to do this, but you can if you want to. (Labeling is explained in Chapter 14.)

3. **Press [GRAPH][WINDOW][6] to select the Polygon tool from the DRAW menu and use this tool to construct the smaller polygons contained in the polygon having intersecting sides.**

   Using the **Polygon** tool is explained in Chapter 14. In the second picture in Figure 15-3, I used the polygon tool to construct the triangles ABX and CDX.

4. **Press [GRAPH][ZOOM][2] to select the Area tool from the MEAS menu. Find the areas of the smaller polygons. Add these areas to find the area of the original polygon.**

The steps in the previous section tell you how to find the area of a polygon. The third picture in Figure 15-3 shows the areas of the smaller triangles contained in the original polygon. The "Performing Calculations" section, later in this chapter, tells you how to add these areas.

Finding the area of the intersection of two polygons

GeoMaster can't find the area of a polygon formed by the intersection of two other polygons because GeoMaster doesn't know that it's there. For example, in the first picture in Figure 15-4, GeoMaster doesn't recognize the existence of the quadrilateral formed by the intersection of the hexagon and triangle because GeoMaster didn't construct that quadrilateral.

To get GeoMaster to find the area of the intersection, you must use GeoMaster to define the polygon formed by this intersection. To do this, you first find the points of intersection of the sides of the original two polygons. And then you use these points to get GeoMaster to define the polygon formed by the intersection. But this is not as easy as it seems because GeoMaster is not capable of finding the points of intersection of two polygons.

However, the steps in the previous subsection on how to find the area of a polygon having intersecting sides can also be followed to find the area of the intersection of two polygons. These steps are briefly listed in the following list. For a more detailed explanation, see the steps in the previous subsection.

1. **Use the Segment tool ([GRAPH][WINDOW][3]) to construct segments over the intersecting sides of the polygons.**

   In the first picture in Figure 15-4, I used the **Segment** tool to construct segments AB, AC, PQ, and QR.

2. **Use the Intersection tool ([GRAPH][WINDOW][ALPHA][PRGM]) to find the points of intersection of the segments constructed in Step 1.**

   In the second picture of Figure 15-4, I used the **Intersection** tool to construct the points of intersection X and Y.

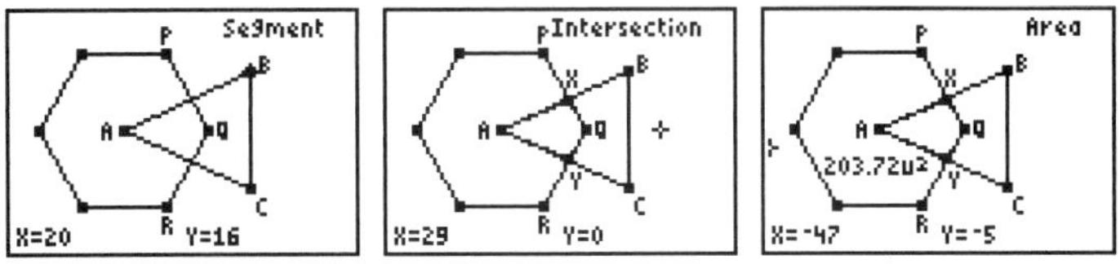

Construct segments   Find intersections   Construct polygon; find area

Figure 15-4: Finding the area of the intersection of two polygons.

3. **Use the Polynomial tool ([GRAPH][WINDOW][6]) to construct the polygon formed by the intersection of the original polygons.**

   In the third picture of Figure 15-4, I used the **Polygon** tool to construct polygon AXQY.

4. **Use the Area tool ([GRAPH][ZOOM][2]) to find the area of the polygon formed by the intersection of the original polygons.**

   This is illustrated in the third picture in Figure 15-4.

Measuring an Angle

For GeoMaster to recognize the existence of an angle, that angle must contain three defining points: the vertex and a point on each side of the angle. Also, GeoMaster recognizes a point only if it appears on the screen as a small square, such as the four points defining the polygon in Figure 15-1. If the angle you want to measure does not contain the small squares representing these three defining points, then GeoMaster cannot measure it. But all is not lost. The previous chapter tells you how to establish the point of intersection of two lines and how to place a point on a line. After you have constructed the three points defining your angle, GeoMaster can find its measure.

TIP: GeoMaster measures all angles in degrees — even if the Mode is set to radian measure. Chapter 3 tells you how to convert degrees to radians. This chapter also tells you how to convert degree measures to degrees, minutes, and seconds.

To find the degree measure of an already constructed angle containing three defining points, follow these steps:

1. **Press GRAPH ZOOM 3 to select the Angle tool from the MEAS menu.**

   The tool you selected appears in the upper-right corner of the screen.

2. **Use the arrow keys to move the cursor to an existing point on one side of the angle and press ENTER to anchor that point.**

   GeoMaster uses a small square to indicate that a point exists. If you don't see a small square on the side of the angle, the first paragraph in this section tells you what to do about the situation.

3. **Use the Arrow keys to move the cursor to the vertex of the angle and press ENTER to anchor the vertex.**

   The vertex must always be the second point that defines the angle.

4. **Use the Arrow keys to move the cursor to an existing point on the other side of the angle and press ENTER to anchor that point.**

   The degree measure of the angle appears near the location of the cursor.

5. **Use the arrow keys to move the angle measure to the location of your choice and press ENTER to anchor that location.**

After measuring an angle, the **Angle** tool remains active until you deactivate it. So if you want to measure more angles, repeat Steps 2 through 5. When you are finished using the tool, deactivate it by pressing CLEAR or by selecting another menu item.

Performing Calculations

GeoMaster is capable of adding and subtracting like measures that GeoMaster found using the **Distance/Length**, **Area**, or **Angle** tools in the **MEAS** menu. By *like measures,* I mean it can, for example, add two area measures; it cannot add unlike measures such as an area measure and an angle measure. GeoMaster cannot multiply, divide, or square a number. So if you want to use GeoMaster to verify the Pythagorean Theorem, you're out of luck.

To add or subtract like measures found by the **Distance/ Length**, **Area**, or **Angle** tool, follow these steps:

1. **Press [GRAPH][ZOOM][4] to select the Calculate tool from the MEAS menu.**

   The tool you selected appears in the upper-right corner of the screen, as illustrated in the first picture in Figure 15-5.

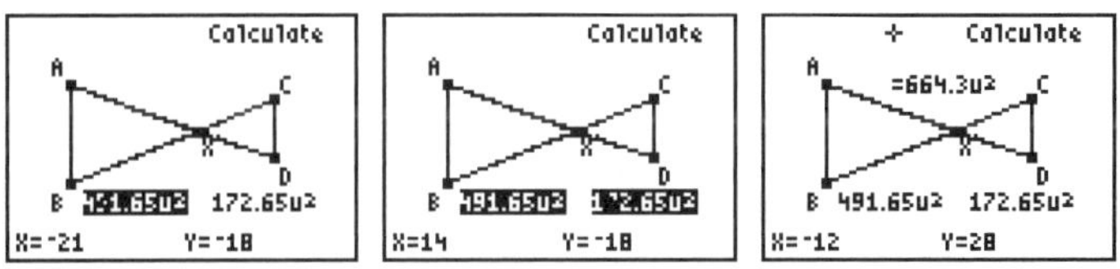

Select area, press [+] Select other area Press [STO▸]

Figure 15-5: Finding the sum of two area measurements.

2. **Use the arrow keys to move the cursor to the first measurement and press [ENTER] to select that measure.**

   Place the cursor at the beginning of a measurement to select it. After you press [ENTER], the measurement is highlighted. This is illustrated in the first picture in Figure 15-5.

3. **Press [+] if you want to add another measurement to the measure selected in Step 2; press [−] if you want to subtract.**

4. **Use the arrow keys to move the cursor to the next measurement in your addition or subtraction problem and press [ENTER].**

   This is illustrated in the second picture in Figure 15-5.

5. **Repeat Steps 3 and 4 to add and/or subtract more measurements.**

   GeoMaster allows you to add and/or subtract up to a maximum of six measurements.

6. **Press [STO▸] to find the answer to the addition and/or subtraction of measurements.**

   This step is not necessary if you added and/or subtracted six measurements. In this case, GeoMaster automatically calculates the answer.

   The calculated answer, preceded by an equal sign, appears on the screen near the location of the cursor.

7. **Use the arrow keys to move the answer to a better location and press [ENTER] to anchor that location. Press [ENTER] again or move the cursor.**

   This is illustrated in the third picture in Figure 15-5.

After adding or subtracting measurements, the **Calculate** tool remains active until you deactivate it. So if you want to add or subtract more measures, repeat Steps 2 through 7. When you are finished using the tool, deactivate it by pressing [CLEAR] or by selecting another menu item.

TIP: The **Calculate** tool will not allow you to use a calculated answer in an addition and/or subtraction problem involving measurements. A calculated answer is preceded by an equal sign.

Finding Slope

The **Slope** tool in the **MEAS** menu can be used to find the slope of a line, segment, ray, vector, or side of a polygon. To find the slope, follow these steps:

1. **Press [GRAPH][ZOOM][5] to select the Slope tool from the MEAS menu.**
2. **Use the arrow keys to move the cursor to the line, segment, ray, vector, or side of a polygon and press [ENTER].**

   The slope appears at the location of the cursor.

3. **Use the arrow keys to move the slope to a better location and press [ENTER] to anchor that location. Press [ENTER] again or move the cursor.**

The **Slope** tool remains active until you deactivate it. So if you want to find the slope of another line or segment, repeat Steps

2 and 3. When you are finished using the tool, deactivate it by pressing [CLEAR] or by selecting another menu item.

Constructing a Vector Sum

To construct the sum of two vectors, follow these steps:

1. **Press [GRAPH][ZOOM][6] to select the Vector Sum tool from the MEAS menu.**
2. **Use the arrow keys to move the cursor to the first vector in the sum and press [ENTER].**

   The selected vector appears as a dashed vector, as illustrated in the first picture in Figure 15-6.

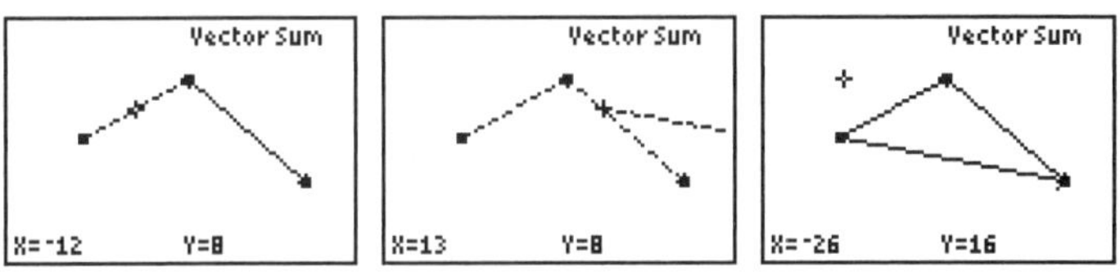

Select first vector | Select second vector | Move vector sum

Figure 15-6: Constructing the sum of two vectors.

3. **Use the arrow keys to move the cursor to the second vector in the sum and press [ENTER].**

   The vector sum is constructed at the location of the cursor, as illustrated in the second picture in Figure 15-6.
4. **Use the arrow keys to move the vector sum to a better location and press [ENTER] to anchor that location. Press [ENTER] again or move the cursor.**

   This is illustrated in the third picture in Figure 15-6.

The **Vector Sum** tool remains active until you deactivate it. So if you want to construct the sum of more vectors, repeat Steps 2 through 4. When you are finished using the tool, deactivate it by pressing [CLEAR] or by selecting another menu item.

Finding the Equation of a Line/Circle or the Coordinates of a Point

The **Eqns/Coords** tool in the **MEAS** menu can be used to find the equation of a line, the equation of a circle, or the coordinates of an already constructed point. It will not find the equation of a segment, ray, vector, or side of a polygon. To find the equation of a line or a circle, or to find the coordinates of a point, follow these steps:

1. **Press [GRAPH][ZOOM][7] to select the Eqns/Coords tool from the MEAS menu.**
2. **Use the arrow keys to move the cursor to the object whose equation or coordinates you want to find and press [ENTER].**

   The equation or coordinates appear at the location of the cursor.
3. **Use the arrow keys to move the equation or coordinates to a better location and press [ENTER] to anchor that location. Press [ENTER] again or move the cursor.**

The **Eqns/Coords** tool remains active until you deactivate it. So if you want to find the equation of other lines or circles, or the coordinates of other points, repeat Steps 2 and 3. When you are finished using the tool, deactivate it by pressing [CLEAR] or by selecting another menu item.

Index

• E •

• F •

• G •

• N •

• O •

• P •

• Q •

• R •

• S •

• T •

Notes

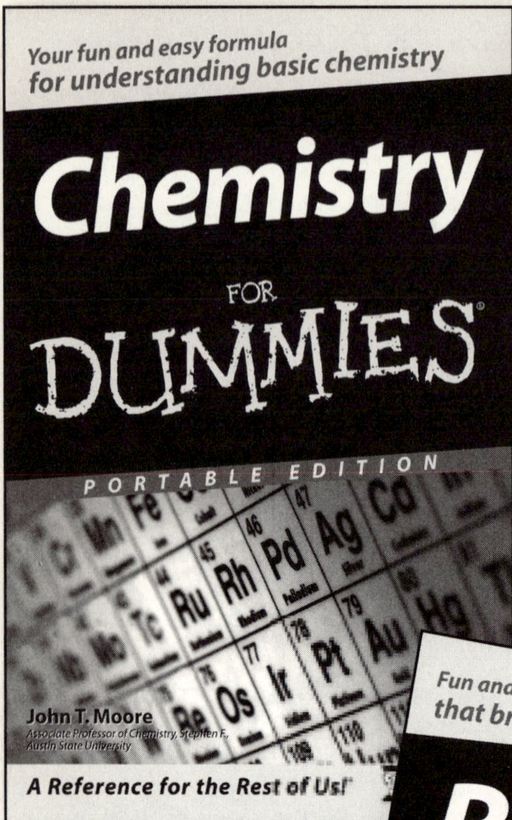
Your fun and easy formula
for understanding basic chemistry
Chemistry
FOR
DUMMIES
PORTABLE EDITION
John T. Moore
Associate Professor of Chemistry, Stephen F. Austin State University
A Reference for the Rest of Us!

Fun and easy guide
that brings biology to life!
Biology
FOR
DUMMIES
PORTABLE EDITION
Donna Rae Siegfried
Author of Anatomy & Physiology For Dummies
A Reference for the Rest of Us!
FREE eTips at dummies.com

The fun and easy way® to
speak and write with confidence
English Grammar
FOR
DUMMIES®
PORTABLE EDITION
Parts of Speech
adjective - describes somet
somebody. ex.→lumpy, soft, ug
adverb - tells how somethin
It modifies a verb and usu
FREE eTips at
dummies.com®
ence for the Rest of Us!®
The fun and easy way® to
start speaking Spanish
Spanish
FOR
DUMMIES®
PORTABLE EDITION
Words to Know
mucho moo-cho a lot; much
difícil dee-fee-seel difficult; hard
fácil fah-seel easy
gustar goos-tahr to like
hablar ah-bvlahr to speak; to talk
este ehs-the this one
otro oh-troh the other one
Susana Wald
Language instructor and translator
A Reference for the Rest of Us!®
FREE eTips at
dummies.com®

Your survival guide to geometry —
with this friendly guide, it's elementary, my dear!
Geometry
FOR
DUMMIES
PORTABLE EDITION
Wendy Arnone, PhD
A Reference for the Rest of Us!

Fun and easy solutions
for math problems in your daily life
Math Basics
FOR
DUMMIES
PORTABLE EDITION
Charles Seiter
A Reference for the Rest of Us!
FREE eTips at
dummies.com

Power up your education
with the right technology!
Connected
on Campus
FOR
DUMMIES®
PORTABLE EDITION
A Reference for the Rest of Us!®
FREE eTips at
dummies.com®

Add up the advantages
of this handy graphing calculator!
TI-83/84
Graphing Calculator
FOR
DUMMIES
PORTABLE EDITION
C.C. Edwards
Author of TI-83 Plus Graphing Calculator For Dummies
A Reference for the Rest

The fun and easy way to
make the most of iPod and iTunes
iPod & iTunes
FOR
DUMMIES
PORTABLE EDITION
Tony Bove
Cheryl Rhodes
Authors of iLife All-in-One Desk Reference For Dummies
FREE eTips at
dummies.com
A Reference for the Rest of Us!